Cambridge Physical Tracts

SEMI-CONDUCTORS & METALS

SEMI-CONDUCTORS & METALS

An introduction to the electron theory of metals

by

A. H. WILSON, M.A.

Fellow of Trinity College, Cambridge

CAMBRIDGE

AT THE UNIVERSITY PRESS

1939

CAMBRIDGE UNIVERSITY PRESS
Cambridge, New York, Melbourne, Madrid, Cape Town,
Singapore, São Paulo, Delhi, Tokyo, Mexico City

Cambridge University Press
The Edinburgh Building, Cambridge CB2 8RU, UK

Published in the United States of America by Cambridge University Press, New York

www.cambridge.org
Information on this title: www.cambridge.org/9780521178594

First published 1939
First paperback edition 2011

A catalogue record for this publication is available from the British Library

ISBN 978-0-521-17859-4 Paperback

GENERAL PREFACE

It is the aim of these tracts to provide authoritative accounts of subjects of topical physical interest written by those actively engaged in research. Each author is encouraged to adopt an individualistic outlook and to write the tract from his own point of view without necessarily making it "complete" by the inclusion of references to all other workers or to all allied subjects; it is hoped that the tracts may present such surveys of subjects as the authors might give in a short course of specialized lectures.

By this means readers will be provided with accounts of those subjects which are advancing so rapidly that a full-length book would be out of place. From time to time it is hoped to issue new editions of tracts dealing with subjects in which the advance is most rapid.

M. L. O.
J. A. R.

CONTENTS

ACKNOWLEDGMENTS

Figures 5, 7, 8, 13, 15, 19, 28 are reproduced from the author's book *The Theory of Metals*, where their sources are noted. Figure 9 is from *The Reviews of Modern Physics*; figures 31–37 from *The Proceedings of the Cambridge Philosophical Society*; and figures 21 and 22 from *The Philosophical Magazine*.

PREFACE

My object in writing this tract has been to give a simplified account of some of the main achievements of the theory of metals in the last ten years. The difficulties and dangers of presenting such a simplified and condensed account of a highly complex subject are so obvious and so great that I should have hesitated to attempt to write the tract were it not that there is now a sufficient number of treatises on the subject to which the reader can refer for further information.

The method of presentation adopted is to try to make clear the physical principles on which the theory is based and to derive the results wherever possible by simplified arguments. The book therefore attempts to give more than a superficial account, but I hope that no one will be misled into assessing the arguments at a higher value than they deserve. When it has not been possible to derive a result by a method substantially easier than that used in the strict mathematical theory as set out in my larger book referred to in the bibliography, I have simply quoted the result. Although the treatment is an elementary one, it cannot be pretended that all parts of the book are easy reading. Some parts of the theory are intrinsically more difficult than others, and all that I can hope is that I have given a readable account of the simpler parts of the subject and supplied some help to those who wish to understand the harder parts. On account of the necessity of keeping the size of the tract within reasonable limits, it has been found impossible to include accounts of the very important subjects of alloys and metal optics.

My thanks are due to Mr J. A. Ratcliffe, who has made many suggestions which have simplified and clarified the exposition.

A. H. W.

November 1938

BIBLIOGRAPHICAL NOTE

The following books deal with the modern theory of metals.

F. Bloch. "Elektronentheorie der Metalle", *Handbuch d. Radiologie*, vol. 6, part 1, 2nd ed. (Leipzig, 1933.)

L. Brillouin. *Die Quantenstatistik.* (Berlin, 1931.)

H. Fröhlich. *Elektronentheorie der Metalle.* (Berlin, 1936.)

N. F. Mott and H. Jones. *The theory of the properties of metals and alloys.* (Oxford, 1936.)

L. Nordheim. "Kinetische Theorie des metallischen Zustandes", *Müller-Pouillets Lehrbuch der Physik*, vol. 4, part 4, 11th ed. (Braunschweig, 1934.)

R. Peierls. "Elektronentheorie der Metalle", *Ergebnisse d. exakt. Naturwiss.* vol. 11. (Berlin, 1932.)

A. Sommerfeld and H. Bethe. "Elektronentheorie der Metalle", *Handbuch d. Physik*, vol. 24, part 2. (Berlin, 1933.)

A. H. Wilson. *The theory of metals.* (Cambridge, 1936.)

A survey of the empirical data is given in

Handbuch d. Metallphysik, vol. 1 (Leipzig, 1935): part 1, "Gitteraufbau metallischer Systeme" by U. Dehlinger, and part 2, "Physikalische Eigenschaften der Metalle" by G. Borelius.

References in the text to formulae taken from my larger book mentioned above are given as *T.M.* p. x, equation (y). References are not in general given to standard results which can easily be found in the text-books. The bibliography at the end of each chapter is confined (1) to less well-known work and work which is so recent that it cannot be found in the text-books, and (2) to books and articles giving further information concerning the subject-matter of the chapter.

Chapter I

FUNDAMENTAL PRINCIPLES

1·1. Introduction.

The modern electron theory of metals is a direct successor of the theory of Drude and Lorentz, which was based on the hypothesis that the electrons in metals are free to move from atom to atom, while those in insulators are not. This fundamental idea of a free electron has been made precise and has been used to elucidate a great many phenomena. Some of the more important applications are indicated below.

Electrons in a solid can move freely from one atom to another owing to the typical quantal effect known as the "tunnel effect" (§ 1·2), but, just as there is a distinctive energy level system in an atom, so the energy levels of the electrons in a solid have a characteristic structure which determines the properties of the substance. Solids in which the electrons form closed groups have properties differing from those of solids in which the electrons form open groups, the former being insulators and the latter being metals (§ 1·3), while if the electrons nearly form a closed group the solid is a semi-metal (§ 2·3). Also, if the number of electrons outside a closed group is zero when the temperature is zero and increases with the temperature, the solid is a semi-conductor (chapter IV).

The cohesive forces in a metal differ considerably from the normal valency forces of classical chemistry, which are based on the idea that a bond is formed when two atoms share a pair of electrons. In a metal the cohesive forces are caused by the interaction of the free electron gas with the metallic ions and are essentially long-range forces which are largely determined by the number of electrons present and not to any great extent by the coordination number (the number of atoms which are nearest neighbours of a given atom) (chapter III). The cohesive forces of alloys are caused partly by the attractions of the different ions and partly by the interaction of the free electrons with the ions. For those alloys in which the latter effect is predominant, the theory shows how the energy level structure affects the cohesion, and it shows that the so-called intermetallic compounds should occur at compositions which bear no relation to the compositions at which compounds would be expected according to the normal valency rules (§ 3·6).

The initial successes of the early theories of metals were rapidly overshadowed by the even greater difficulties which they created, chief among which was the difficulty of the specific heat. According to the classical theory, each free electron should contribute $\frac{3}{2}k$ to the specific heat, but in fact the specific heats of metals and insulators do not show any large difference and can be explained quite well by assuming that the free electrons do not contribute appreciably. This difficulty has been removed by applying Fermi-Dirac statistics instead of Maxwell statistics to the free electrons (§ 1·4). The free electrons make a small contribution to the specific heat, which can only be detected either at very low temperatures where it is larger than the specific heat of the lattice vibrations or at very high temperatures where its temperature variation is larger than that of the normal specific heat (§§ 5·1 and 5·2).

The use of the Fermi-Dirac statistics also enables us to understand why the paramagnetic susceptibility of many metals is independent of the temperature, whereas the susceptibility of a paramagnetic gas is inversely proportional to the temperature (§ 5·3). Further, in ferromagnetic metals the magnetic moment per atom, the magneton number, is determined by the structure of the energy level system, and the theory gives an explanation of the fact that the magneton numbers of actual metals are not integral, nor are they simply related to the number of valency electrons per atom (§ 5·6).

The treatment of conduction problems requires a second fundamental idea, that of the free path. In a perfect crystal an electron can move in a straight line without being deflected (§ 1·2), and so the conductivity is infinite. The resistance in a metal is caused by the scattering of the electrons by irregularities in the crystal; these irregularities may be due to the presence of impurity atoms and strains or to the temperature motion of the atoms (§ 6·21). The conduction of heat and the thermoelectric effects are second order effects which can be calculated when the free path is known (§§ 6·4 and 6·5).

While a complete shell of electrons has no conducting properties, a shell which is nearly complete has certain anomalous properties, chief among which is that the shell behaves as if it had a positive charge (§ 1·22).

This is the explanation of the fact that the Hall coefficients and Thomson coefficients of metals are sometimes negative and sometimes positive (§§ 4·2, 6·3 and 6·5) and it is not necessary to postulate the existence of positive carriers to explain the occurrence of the anomalous positive signs.

The only really unsolved problem is superconductivity, which is so baffling that up to the present we have no idea in which direction to look for an explanation.

Chapter I deals with all the fundamental ideas which are necessary for a rough qualitative survey of the subject, while chapter II is concerned with the further general theory required for discussions of the finer details. The remaining chapters are devoted to the applications of the results set forth in the first two chapters.

1·2. The motion of an electron in a perfect crystal.

Since the problem of determining the electronic states of a solid is of enormous complexity we have to introduce great simplifications in order to obtain any results at all. We assume as a first approximation that each electron can be treated independently. We do not neglect the other electrons entirely; we replace their effects by the average or "smeared" field which they produce, but we do not take into account correlations in the positions of the electrons when taking this average. This assumption amounts to ascribing to each electron a definite state of motion and a definite energy. It is an assumption which has proved very fruitful in the theory of atomic and molecular structure, and, though it probably has less justification in solids, nevertheless it provides the best approach to the problem.

We have now to determine the motion of a single electron in a fixed three-dimensional field which has the periodicity of the crystal. Let us consider for simplicity an electron and a linear chain of N similar equidistant atoms at a distance a apart. If a is large, the electron will move in an orbit round one particular atom uninfluenced by the presence of the other atoms. If, however, a is of the order of the interatomic distances actually found in crystals, the electron will not remain attached to one atom, even when its kinetic energy is not sufficient to carry it over the potential hump between two atoms. For, if the electron were placed in a similar orbit on any other atom, it would be in a state with the same energy, and, according to quantum mechanics, an electron can always jump from one state to a state of equal energy. The probability of such a jump decreases exponentially as the height and the breadth of the potential hump are increased, and it is therefore only appreciable when the atoms are near together and when the electron is not too tightly bound. The apt phrase "the tunnel effect" has been coined to describe the leaking of electrons through potential barriers.

From the preceding discussion we see that, in a stationary state, an electron in a perfect crystal does not revolve round one atom, but it moves steadily through the crystal jumping from one atom to the next. If, however, the lattice is not perfect, the motion is more complicated. Still considering a linear chain, let us suppose that an atom B is displaced so that it is nearer to its neighbour A than to its neighbour C. Then, if an electron moving through the crystal in the direction ABC arrives at B, the probability that it should jump to C is less than normal since the distance BC, and therefore the potential hump between B and C, is greater than the normal one. There is, therefore, a probability that the electron will jump back to A and be reflected instead of continuing its course. A similar state of affairs occurs in a real three-dimensional crystal, except that the electron can be deflected by an irregularity through any angle and not just reflected back along its original path.

The irregularities may be caused by the presence of foreign atoms, by strains, or by the temperature motion of the lattice. When the irregularities are large, as in liquid metals, the electrons can move from atom to atom but the number of deflexions suffered per second is much larger than in solids, and therefore liquid metals like mercury are distinctly poor conductors. It is only in a perfect lattice, at absolute zero temperature, that an electron can continue to move through a large distance without being deflected.

Let us consider in more detail the possible energy levels of the linear chain. As a first approximation we assume that the electron is revolving in the ground orbit round some particular atom and that the electrostatic forces of the other atoms can be neglected. This state is degenerate since the electron could equally well be in a similar orbit round any of the $N-1$ other atoms. If we now introduce the electrostatic forces which have been neglected, this degenerate state splits up into N non-degenerate states lying very close together, each of which represents a state of motion in which the electron is no longer localized but moves freely through the crystal. In the limit, when N becomes infinite, these energy levels form a continuum or band whose width is proportional to the electrostatic forces between the atoms. When the band is wide the electrostatic forces are large and the mobility of the electron is large. When the band is narrow the probability of the electron jumping from one atom to the next is small.

Instead of considering the electron to be revolving in the ground state round one atom, we could have considered it to be in an excited state. This state would also give rise to a band of N energy levels. We therefore see that the energy spectrum consists of a set of continuous bands separated by intervals for which no energy levels are possible. Each of these bands contains as many energy levels as there are atoms in the crystal. So long as the width of the bands is less than the energy difference between the levels of the free atom it is possible to correlate the bands with the atomic levels from which they are derived. When the width is larger than these energy differences, which is always true for the highly excited levels, a unique correlation is no longer possible. The regions of disallowed energies still exist, however, but they get narrower as the energy increases. The general form of the potential energy of an electron and the arrangement of the energy bands are shown schematically in fig. 1.

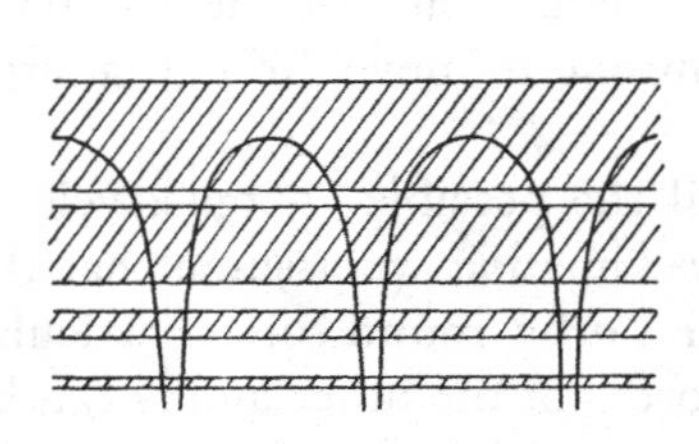

Fig. 1. Energy bands (shaded) in a one-dimensional lattice.

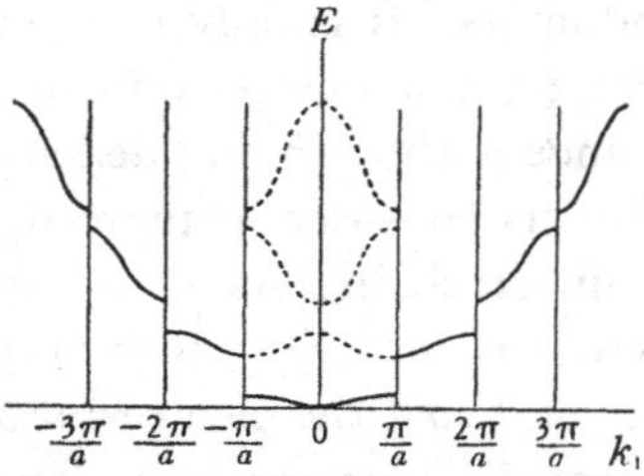

Fig. 2. The energy as a function of the wave number. The dotted curves give the energy in terms of the reduced wave vector.

1·21. The preceding discussion really assumes that the electrons are tightly bound, i.e. that the electrostatic forces are such that the band width is smaller than the energy difference between the atomic levels. When this is not true we can approach the problem from another angle. When the crystal is highly compressed (or if we consider highly excited states) the kinetic energy of the electrons is large, and the potential humps are narrow and comparatively unimportant. Therefore, as a first approximation, we treat the electron as if it were perfectly free; it is then represented by a plane wave moving through the crystal. The wave function is

$$\psi = e^{i(k_1x+k_2y+k_3z)}, \tag{1·1}$$

the energy is

$$E = \frac{h^2}{8\pi^2 m}(k_1^2+k_2^2+k_3^2) \tag{1·2}$$

and the momentum is $h(k_1, k_2, k_3)/2\pi$. The effect of the three-dimensional periodic field is to introduce modulations into ψ

with the same period as the field. It can be shown (*T.M.* p. 38, equation (61)) that ψ has the form

$$\psi = e^{i(k_1x+k_2y+k_3z)}\, u_{k_1,k_2,k_3}(x, y, z), \tag{1·3}$$

where $u(x, y, z)$ is periodic in the unit cell of the crystal. So long as we consider only slow electrons with long wave-lengths so that k_1, k_2 and k_3 are small compared with $1/a$ (a is the lattice constant), the modulations do not have very much effect on E. When, however, k_1, say, is nearly equal to π/a, resonance sets in and the energy spectrum is radically altered. The electron is strongly reflected by the lattice planes and a standing wave is set up. These reflexions are the exact analogues of the Bragg reflexions of X-rays. They occur whenever the electron has the appropriate wave-length and direction of motion for reflexion by a particular set of lattice planes.

It is not possible to see the effect of the reflexions on the energy spectrum, and to determine the behaviour of the energy as a function of the wave vector (k_1, k_2, k_3), without calculation. What actually happens is shown in fig. 2, p. 6; wherever there is a reflexion the energy has a discontinuity. In the one-dimensional case the discontinuities occur when $k_1 = \pm n\pi/a$ (n a positive integer), and the energy levels for which $\pm k_1$ lies between $n\pi/a$ and $(n-1)\pi/a$ form a band or zone separated from the neighbouring bands by a region in which there are no energy levels. (Positive values of k_1 correspond to electrons moving to the right, say, and negative values to electrons moving to the left.) These bands are, of course, the same as the bands discussed at the beginning of this section. Both methods of approach give the same qualitative information about the energy spectrum.

It is often convenient to restrict k_1, k_2, k_3 to lie between $\pm\pi/a$, since then the different parts of a zone are joined up. To do this we have to subtract from k_1, k_2, k_3 suitable multiples of π/a. The factors such as $e^{2\pi n x/a}$ which are left over can be absorbed into $u(x, y, z)$ since they are periodic in the lattice constant. In this case the wave vector $\mathbf{k}$ is called the reduced wave vector.

The energy ranges for which no steady motion of the electron exists are analogous to the frequency ranges near an absorption line for which anomalous dispersion of light occurs. According to the theory of dispersion, there is a range of frequencies round any resonance frequency of a substance for which a light wave cannot be propagated at all if damping is neglected, while, if damping is taken into account, the light wave can exist but it is strongly absorbed. Correspondingly, an electron impinging on a metal with an energy which lies in a forbidden region can enter the metal but it is strongly reflected, whereas if its energy lies in an allowed band it can pass freely through the metal. The well-known experiments of Davisson and Germer show very clearly this anomalous dispersion of electrons.

In a real three-dimensional crystal the zone structure can be very complicated. The details are discussed in chapters II, III and IV.

1·22. The velocity components v_1, v_2, v_3 of an electron are given in terms of the derivatives of E by de Broglie's relation. Thus

$$v_1 = \frac{2\pi}{h}\frac{\partial E}{\partial k_1}. \qquad (1\cdot4)$$

The form of v_1 is shown in fig. 3. When $k_1=0$, $v_1=0$, and, as k_1 increases, v_1 increases, reaches a maximum and finally becomes zero again at the top of the band.

Electrons for which v_1 lies between the maximum and the minimum behave normally, while those which lie to the right of the maximum or to the left of the minimum behave abnormally, since when the energy increases the velocity decreases in absolute magnitude. To see the importance of this, consider the action of an electric field $\mathscr{E}$ which is along the x axis. The charge of the electron being $-\epsilon$, the increase in the kinetic energy of the electron per second is $-\epsilon\mathscr{E}v_1$. Hence, by (1·4),

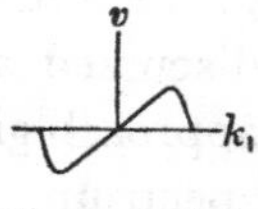

Fig. 3. The velocity as a function of the wave number.

$$-\epsilon\mathscr{E}v_1 = \frac{dE}{dt} = \frac{\partial E}{\partial k_1}\frac{dk_1}{dt} = \frac{h}{2\pi}v_1\frac{dk_1}{dt},$$

and so
$$\frac{dk_1}{dt} = -\frac{2\pi}{h}\epsilon\mathscr{E}. \qquad (1\cdot5)$$

Further, the acceleration is given by
$$\frac{dv_1}{dt} = \frac{d}{dt}\frac{2\pi}{h}\frac{\partial E}{\partial k_1} = \frac{2\pi}{h}\frac{\partial^2 E}{\partial k_1^2}\frac{dk_1}{dt}$$
$$= -\frac{4\pi^2}{h^2}\epsilon\mathscr{E}\frac{\partial^2 E}{\partial k_1^2}. \qquad (1\cdot6)$$

When the electrons are perfectly free and
$$E = h^2(k_1^2 + k_2^2 + k_3^2)/(8\pi^2 m),$$
equation (1·6) becomes
$$\frac{dv_1}{dt} = -\frac{\epsilon\mathscr{E}}{m},$$
which is the ordinary acceleration equation for a free particle. However, as fig. 2 shows, $\partial^2 E/\partial k_1^2$ can take on negative values, and, when it does so, the acceleration is in the opposite direction to the normal one. For states sufficiently near the top of a band we can write $E = A - h^2(k_1^2 + k_2^2 + k_3^2)/(8\pi^2 m^*)$, since the energy is a maximum at the top of the band, k_1, k_2 and k_3 being measured from the position of the maximum. In this case
$$\frac{dv_1}{dt} = \frac{\epsilon\mathscr{E}}{m^*},$$
and thus we can say that the electron behaves as if it had either a negative mass or a positive charge. (The second statement is the more usual one.) m^* is called the effective mass. We therefore have the very important result that an electron behaves as if it has a positive charge if it occupies an energy level lying in the region in which the energy curve as a function of the wave number is concave downwards, i.e. near the top of an energy band. This peculiarity has no effect on the electrical conductivity since the conductivity is proportional to the square of the charge. It only plays an important role for those phenomena, such as the Hall and thermoelectric effects, which are proportional to the first power of the charge. This is discussed further in § 1·3.

The energy curves of tightly bound electrons are similar

qualitatively to those of nearly free electrons, but the energy bands are narrower and the anomalous regions in which $\partial^2 E/\partial k_1^2$ is negative are more marked. It is no longer possible to represent E as a quadratic function of k_1, k_2 and k_3 over a considerable range, but this can of course be done when we are only interested in a small range of the wave vector, as happens, for example, in the theory of electrical conductivity. We may then write instead of (1·2)

$$E = \frac{h^2}{8\pi^2 m^*}(k_1^2 + k_2^2 + k_3^2) + \text{constant}. \qquad (1{\cdot}7)$$

The effective mass m^* can be considerably larger than m if the potential humps between the atoms are large. It can also be negative, but in this case we prefer to keep to positive masses and treat the charge as positive instead.

1·3. The exclusion principle: metals and insulators.

Schrödinger's equation governs the motion of the electrons in a crystal, but the possible configurations are further limited by Pauli's exclusion principle which states that not more than two electrons can occupy the same energy level, and, if there are two, they must have their spins in opposite directions. This enables us to explain why it is possible to divide solids into the two distinct classes of metals and insulators, even although the electrons can move freely through the crystal.

Consider the behaviour of a solid at the absolute zero, and consider for the moment only the valency electrons. The state of lowest energy for an electron is the state of zero kinetic energy. According to the classical theory all the valency electrons would be in this state, but this is not possible on account of the exclusion principle. Instead the states, starting from the lowest, are successively filled by two electrons with opposite spins until all the electrons are accommodated.

Consider first the case in which the number of valency electrons per atom is such as just to fill an energy band. The effect of an external electric field is to tend to produce a current by accelerating the electrons moving in one direction and retarding those moving in the opposite direction, or in other words to increase the energy of the electrons moving in one direction and to decrease the energy of the others. In the case we are considering this is impossible since there are no vacant energy levels to which the electrons can be transferred. Of course there are the energy levels of the next allowed band, but they lie about 1 e.volt higher, so a field of the order of at least 10^6 volts/cm. would be required to produce a transition. In this case, therefore, the solid behaves as an insulator.

Next suppose that the number of electrons present is not sufficient to fill an energy band, as would happen, for example, for a linear chain in which there was one valency electron per atom. So far as most of the electrons are concerned, an electric field is just as ineffective as in the previous case, but for the most energetic electrons the situation is entirely different. For these electrons there are now plenty of vacant levels into which they can be transferred by an external electric field, and in a perfect lattice they would be accelerated by the field until their energies reached the limit of the allowed band. In actual fact, however, the accelerating effect of the field is resisted by scattering agents such as foreign atoms, distortions and the thermal motion of the lattice, so that the resultant change in the distribution of the electrons over the energy levels is small. The important point is that, when the electrons do not fill an energy band, a redistribution of the electrons over the energy levels is possible and a current can be set up by an electric field. The solid is, therefore, a metal.

When a band is nearly full it has anomalous properties since the most energetic electrons, which alone contribute to the conductivity, behave as if they had a positive charge (§ 1·22). Since a completely filled band does not conduct at all, it is often convenient to say that the vacant spaces or "holes" in a nearly full band are responsible for the current, the motion of an electron in one direction being equivalent to the motion in the opposite direction of the vacant space which it leaves behind. This interpretation must, however, be used with caution since it is always the electrons which really carry the current and which are accelerated by external fields. Nearly complete bands behave anomalously because the most energetic electrons occupy energy levels for which $\partial^2 E/\partial k_1^2$ is negative. When there are sufficient holes for the energy levels to fall in the region where $\partial^2 E/\partial k_1^2$ is positive, the electronic distribution behaves normally. When we talk about "holes" in discussing semi-conductors in § 4·2 and the Hall effect in § 6·31, we always assume that the number of "holes" is so small that the anomalous behaviour occurs, and we often describe them loosely as "positive holes".

For a linear lattice, two electrons per atom suffice to fill an energy band, but this is not so for a three-dimensional crystal. A detailed knowledge of the zone structure is necessary before we can say whether a given solid will be a metal or not; this is discussed in § 2·3. We now see that it is immaterial whether we regard the inner electrons as being mobile or not; in any case they cannot contribute to the current since they belong to completely filled bands. However, we prefer to regard the inner electrons as "atomic electrons", definitely attached to one atom, and to reserve the idea of "free" or "metallic electrons" for the most loosely bound electrons.

To sum up, although the valency electrons can be considered as being free to move through the lattice, not every solid is a metal, on account of the possibility of closed groups of electrons being formed. A solid is a metal if and only if the valency electrons form an open group.

1·4. The Fermi-Dirac statistics.

The difficulty of the specific heat which proved so fatal to the theory of Drude and Lorentz disappears when proper account is taken of the exclusion principle. According to the classical theory all the free electrons are in the lowest state, i.e. they have zero translational energy, at absolute zero temperature. As the temperature is raised, the electrons acquire kinetic energy and are responsible for a large specific heat. This is obviously wrong, since, if the argument is carried to its logical conclusion, we ought to ascribe the classical specific heat of $\frac{3}{2}k$ to all the inner, core electrons as well; we have no valid reason for applying quantum principles to the core electrons and classical principles to the valency electrons.

In the preceding section we saw that in fact the valency electrons occupy, at the absolute zero, the energy levels ranging from zero energy to a maximum energy, which we call E_0 and which depends, amongst other things, on the number of electrons present. If the temperature is T instead of zero, the distribution is modified, and this modification determines the specific heat. The effect of increasing the temperature is to tend to increase the energy of the electrons, but we must bear in mind the fact that each energy level can accommodate only two electrons with opposite spins. The thermal energy is of the order kT, and so if $E_0 \gg kT$ the electrons with low energies cannot be excited, since the energy levels within reach are already occupied. Only for those electrons with energies near E_0 can the distribution be effectively modified.

1·41. Let $f_0(E)$ be the distribution function, i.e. $f_0(E)$ is the probable number of electrons with a given spin in a state with energy E when the temperature is T. Then $f_0(E)$, which is called the Fermi-Dirac function, must have the following properties.† For $T=0$, $f_0(E)$ is 1 if E lies between 0 and E_0, and $f_0(E)$ is 0 if E is greater than E_0. For $T \neq 0$, $f_0(E)$ must be constant and equal to 1 for small values of E; as E approaches E_0, $f_0(E)$ must begin to decrease and fall practically to zero in an energy interval round E_0 of the order of kT. Thus, instead of $f_0(E)$ dropping sharply to zero at E_0, the fall is rounded off and $f_0(E)$ has a "Maxwellian tail". The actual form of $f_0(E)$ is

$$f_0(E) = \frac{1}{e^{(E-\zeta)/kT}+1}. \qquad (1·8)$$

The quantity ζ is the thermodynamic potential of the electrons, and, when $T=0$, it is the same as E_0. When $\zeta \gg kT$, the electron gas is said to be degenerate, and $T_0 = E_0/k$ is called the degeneracy temperature, since for $T \ll T_0$ the electron gas is degenerate

† Note that $f_0(E)$ is the probability of a given state being occupied; $f_0(E)\,dE$ is *not* the average number of electrons with energies lying in the range dE, since there are many states with energies lying in this range. See § 2·5.

while for $T \gg T_0$ the electron gas behaves classically. The general form of $f_0(E)$ for a degenerate gas is shown in fig. 4.

The order of magnitude of T_0 can be calculated as follows. Let n be the number of electrons per unit volume. Then, in order to apply classical statistics, it is necessary to be able to define the position of an electron with an error $(\Delta x)^3$ which is such that

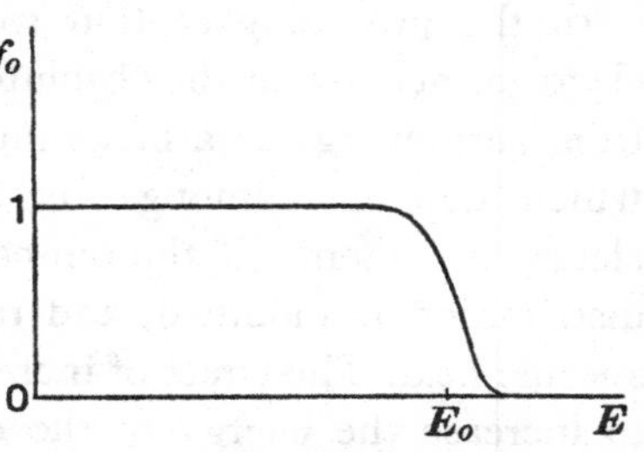

Fig. 4. The Fermi function.

$$\frac{1}{n} \gg (\Delta x)^3, \tag{1·9}$$

so that the electrons can be treated as distinguishable. The corresponding error Δp in the linear momentum is connected with Δx by the relation

$$\Delta x \Delta p \sim h. \tag{1·10}$$

Further, Δp must be small compared with the mean momentum of thermal agitation. Hence

$$(mkT)^{\frac{1}{2}} \gg \Delta p. \tag{1·11}$$

Thus, combining (1·9), (1·10) and (1·11), we see that the condition for the validity of classical statistics is

$$n^{-\frac{1}{3}}(mkT)^{\frac{1}{2}} \gg h.$$

Hence T_0 must be of the order

$$\frac{n^{\frac{2}{3}}h^2}{mk}. \tag{1·12}$$

The exact formula is (*T.M.* p. 16, equation (29))

$$E_0 = kT_0 = \frac{h^2}{8m}\left(\frac{3n}{\pi}\right)^{\frac{2}{3}}, \tag{1·13}$$

a proof of which is given on p. 34. If we put $n = 5·9 \times 10^{22}$, which is the value for silver if we assume that there is one free electron per atom, we find $T_0 = 6 \times 10^{4}\,^\circ$ K. Thus for most purposes we can assume that the electron gas is highly degenerate.

The quantity ζ is not a constant but varies with the temperature. For temperatures which are small compared with T_0, the variation of ζ is small and we can put $\zeta = E_0$. For high temperatures, however, ζ decreases and the Fermi function becomes very spread out. When $T \gg T_0$, ζ actually becomes negative, the Fermi statistics becomes the ordinary classical statistics and nothing is left of $f_0(E)$ but the Maxwellian tail, the curve being concave upwards everywhere.

1·42. Since the drop in the Fermi function is spread over an energy range of the order of kT, while the energy spread of the occupied states is of the order E_0, only a fraction kT/E_0 of the electrons contributes to the specific heat. Thus, although the number of free electrons is large, the effective number contributing to the specific heat is small and there is no great difference between the heat capacity of metals and insulators.

Any quantity, such as the specific heat, which vanishes when we assume the electron gas to be completely degenerate, i.e. when we replace $f_0(E)$ by its value when $T=0$, is called a second order quantity. For such quantities, only those electrons in an energy interval of the order kT round E_0 play a part, and in order to calculate them we need to know the density of states $\mathfrak{n}(E_0)$, the definition of which is that $\mathfrak{n}(E)\,dE$ is the number of energy levels in the range dE. Alternatively, we can use the experimentally measured quantities to determine $\mathfrak{n}(E)$, and that is all the information we can obtain from the experiments.

The calculation of first order quantities† is simple. All we have to do is to replace $f_0(E)$ by 1 if $E < E_0$ and by 0 if $E > E_0$. Second order quantities have to be calculated by using the approximate formula (*T.M.* p. 15, equation (28))

$$-\int_0^\infty \phi(E)\frac{\partial f_0}{\partial E}\,dE = \phi(\zeta) + \frac{\pi^2 k^2 T^2}{3\zeta}\left(\frac{d^2\phi}{dE^2}\right)_{E=\zeta}. \quad (1·14)$$

The order of magnitude of a second order quantity is, however, usually obtainable by elementary arguments.

At first sight we might expect all second order quantities to be determined by the behaviour of the fastest electrons only, while for first order quantities all the electrons would play a part.

† First order quantities are those which are not second order.

This is not so. The electrical resistance and the paramagnetic susceptibility are examples of first order quantities to which only the fastest electrons contribute. In both these cases the entire effect is due to the modification of the electronic distribution by an external field, and the electrons which occupy the low energy levels are ineffective since the neighbouring levels are full.

Chapter II

THE ENERGY LEVELS OF A THREE-DIMENSIONAL CRYSTAL

2·1. Brillouin zones.

The energy spectrum of a one-dimensional lattice, discussed in § 1·2, consists of a number of bands separated by regions in which there are no energy levels, and each band can accommodate exactly two electrons per atom. This simple structure of the energy bands does not occur in real three-dimensional crystals, and the band structure varies widely from metal to metal and determines their characteristic properties.

The fundamental problem is to determine the energy of a state as a function of the wave vector $\mathbf{k}=(k_1, k_2, k_3)$. This can only be done approximately by long and complicated numerical calculations (see § 2·4), but many important qualitative results can be obtained in other ways. For a one-dimensional lattice the energy is a function of one quantum number only, say k_1, and has the form shown in fig. 2, p. 6, discontinuities occurring when $k_1 = \pm n\pi/a$. For three dimensions the energy levels are not so easily visualized. The simplest method is to consider a three-dimensional space, the $\mathbf{k}$ space, in which k_1, k_2, k_3 are rectangular coordinates, and to construct the surfaces of constant energy. For perfectly free electrons E is given by (1·2) and the energy surfaces are spheres. In the general case, if we take one particular direction in the $\mathbf{k}$ space, the energy has the form shown in fig. 2. The general form of the energy contours can therefore be constructed by fitting together sections of this type.

We now consider the discontinuities in the energy more thoroughly. These discontinuities always occur whenever the direction and magnitude of the wave vector of an electron is such that the electron suffers a Bragg reflexion. If the spacing between a particular set of lattice planes is d, Bragg reflexions occur when $n\lambda = 2d \cos\theta$, where n is an integer, λ is the wave length, and θ is the angle between the initial direction of motion and the normal to the reflecting planes. Now $\lambda = 2\pi/k$, where $k = |\mathbf{k}|$, and $k \cos\theta$ is k_n, the component of the wave vector along the normal to the reflecting planes. Thus the condition for Bragg reflexion is*

$$k_n = n\pi/d, \tag{2·1}$$

and it is for these values of $\mathbf{k}$ that the discontinuities in the energy occur. The equation (2·1) represents a plane in the $\mathbf{k}$ space, parallel to the reflecting planes of the crystal, at a distance $n\pi/d$ from the origin. If we consider all possible reflecting planes and all possible orders of reflexion, we obtain a set of planes in the $\mathbf{k}$ space, which divide it up into zones, called Brillouin zones after L. Brillouin who first studied their structure, which are the three-dimensional generalizations of the one-dimensional energy bands discussed in § 1·2. Inside each zone the energy is a continuous function of (k_1, k_2, k_3) and it increases discontinuously when a boundary is crossed in the direction of increasing k.

2·2. The zones for a simple cubic lattice.

Although no metal has a simple cubic structure, the zones for such a lattice serve to illustrate the theory in an elementary manner. Further, the results are needed for the discussion of those cubic lattices which actually occur.

The reflecting planes which have the largest spacing, and so give rise to the smallest values of k_n, are those perpendicular to the crystal axes. The direction ratios of the normals to these planes are $(\pm 1, 0, 0)$ or $(0, \pm 1, 0)$ or $(0, 0, \pm 1)$, and we have $d = a$, where a is the lattice constant. The first zone $(n = 1)$ is therefore the cube bounded by the planes

$$k_1 = \pm\pi/a, \quad k_2 = \pm\pi/a, \quad k_3 = \pm\pi/a. \tag{2·2}$$

The next set of reflecting planes consists of those with the

* To conform to custom we use n in two different senses in (2·1) The suffix n denotes the normal component, while the n on the right-hand side is an integer.

direction ratios $(\pm 1, \pm 1, 0)$ or $(\pm 1, 0, \pm 1)$ or $(0, \pm 1, \pm 1)$, and for these $d = a/\sqrt{2}$. The corresponding zone boundaries (with $n = 1$) are

$$\pm k_1 \pm k_2 = 2\pi/a, \quad \pm k_1 \pm k_3 = 2\pi/a, \quad \pm k_2 \pm k_3 = 2\pi/a, \quad (2\cdot3)$$

and they form the dodecahedron shown in fig. 5 (ii). The second zone lies between this dodecahedron and the fundamental cube (2·2). The third zone consists of the space, lying outside the dodecahedron, defined by the intersections of the planes (2·2) and (2·3); it is shown in fig. 5 (iii). The external

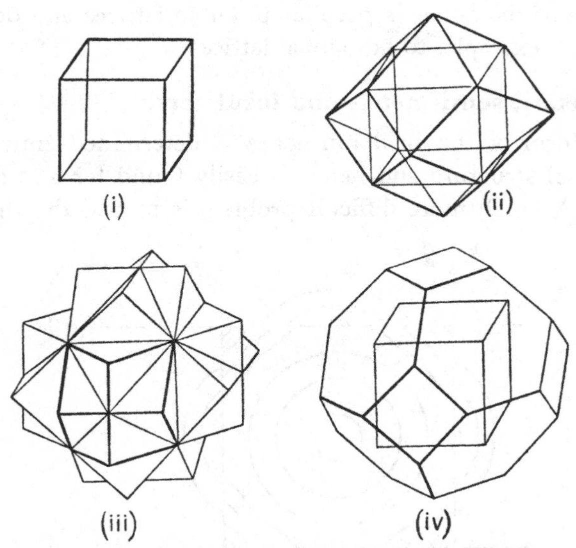

Fig. 5. The first four Brillouin zones of a cubic lattice.

boundaries of the fourth zone are formed by two sets of planes, (i) the planes with direction ratios $(\pm 1, \pm 1, \pm 1)$ and with $d = a/\sqrt{3}$, $n = 1$, (ii) the planes with direction ratios such as $(1, 0, 0)$ with $n = 2$. They are therefore

$$\pm k_1 \pm k_2 \pm k_3 = 3\pi/a \quad (2\cdot4)$$

and $\quad k_1 = \pm 2\pi/a, \quad k_2 = \pm 2\pi/a, \quad k_3 = \pm 2\pi/a. \quad (2\cdot5)$

The surface of this zone, consisting of eight hexagons and six squares, is shown in fig. 5 (iv).

The fundamental cube defined by (2·2) contains as many energy levels as there are atoms in the crystal, and it can therefore accommodate exactly two electrons per atom, with opposite spins. The number of levels in the other zones is proportional to the volume they occupy in **k** space; this has to be found by mensuration. It is found that each zone has the same volume as the fundamental cube. Therefore the dodecahedron (2·3) can accommodate exactly four electrons per atom, and the truncated octahedron defined by (2·4) and (2·5) can accommodate eight electrons per atom. The equality of the volumes of the zones is peculiar to cubic lattices and does not apply, for example, to hexagonal lattices.

2·3. Metals, semi-metals and insulators.

The form of the Brillouin zones is determined entirely by the crystal structure and hence is easily found for any type of lattice. A much more difficult problem is to find the shape of

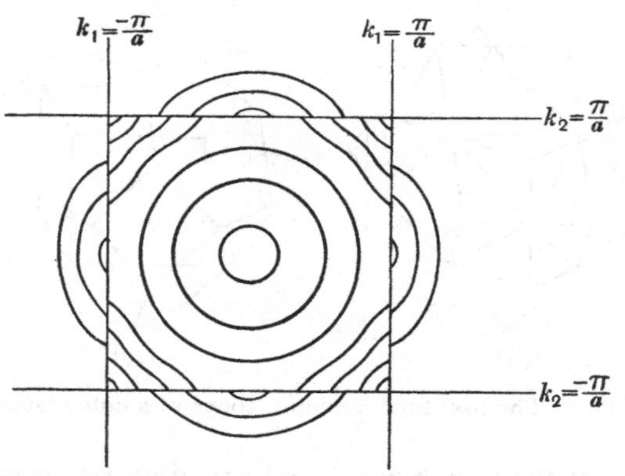

Fig. 6. The energy contours in the plane $k_3=0$.

the energy contours and the magnitude of the energy discontinuities. Since these depend on the electrostatic forces, they can be found only by actually solving the Schrödinger equation for the crystal. Before we consider the results of these calculations, it is convenient to consider the possible behaviour of the energy contours and their effect on the properties of the crystal.

A section of the energy contours by the plane $k_3 = 0$ is shown schematically for a simple cubic lattice in fig. 6; the contours are given for the whole of the first zone, but only for a few of the lowest energy levels in the second zone. Now, although the energy of a state in the second zone is greater than the energy of the adjacent state in the first zone, we cannot conclude that all the energy levels of the first zone are lower than those of the second zone. For, when the energy contours are as shown in fig. 6, the states with the greatest energy in the first zone are given by $k_1 = k_2 = k_3 = \pm\pi/a$, while the states with the least energy in the second zone are given by $k_1 = \pm\pi/a$, $k_2 = k_3 = 0$, and four other states obtained by permuting k_1, k_2, k_3. All that we know with certainty is that

$$E_2(\pi/a, 0, 0) > E_1(\pi/a, 0, 0),$$

where the suffixes denote the zones, and, since

$$E_1(\pi/a, \pi/a, \pi/a) > E_1(\pi/a, 0, 0),$$

we cannot say without calculation which of $E_1(\pi/a, \pi/a, \pi/a)$ and $E_2(\pi/a, 0, 0)$ is the greater.

We distinguish two cases:

(1) All the energy levels of the second zone lie above the energy levels of the first zone, i.e. the energy discontinuities are large. In this case all the energy levels of the first zone must be occupied (at the absolute zero) before any electrons go into the second zone, and if the number of electrons present is just sufficient to fill the first zone they will do so. This type of crystal is an insulator.

(2) Some of the energy levels of the second zone lie lower than the highest levels of the first zone. In this case the first zone can never be full of electrons while the second zone is completely empty. If there are just enough electrons present to fill the first zone, they will not do so, but will occupy some of the levels of the second zone and leave some vacant levels in the first. Thus although the number of electrons present may be sufficient to make it possible for the solid to be an insulator, yet the energy gaps may be so small that the solid is in fact a metal. There is another case (2*a*) which cannot be sharply

distinguished from the preceding case. If the energy gaps are neither too large nor too small, it may happen that the energy levels of the two zones only just overlap. In this case, if there are just enough electrons present to fill the first zone, there will be a small number of vacant places in the first zone and the same number of electrons in the second zone. The solid must still be a metal, but the number of effective free electrons will be so small that the solid will be a poor or semi-metal, such as bismuth.

2·4. Methods for the calculation of energy levels.

In order to discuss whether the metallic or the gaseous state is the more stable, it is necessary to know the energy of a state not only as a function of **k** but also of a, the lattice constant. A method of finding the energy for the lowest state, i.e. for $\mathbf{k}=0$, was given by Wigner and Seitz (1). The wave function $u_0(x, y, z)$ of this state has the lattice constant for its period (see equation (1·3)), and by symmetry the component of grad u_0 along the lines joining the centres of two neighbouring atoms must vanish half way between them. Consider in particular sodium, which has a body-centred cubic lattice, and fix attention on an atom at the centre of one particular cell. The lines joining it to its nearest neighbours are the eight lines to the corners of the cube and the six lines to the centres of the neighbouring cells. If we draw planes bisecting these fourteen lines at right angles, we obtain a polyhedron which is the truncated octahedron shown in fig. 5 (iv), p. 19. Now the wave function u_0 must be such that it and its derivatives are continuous and the normal component of grad u_0 must vanish at the centres of the fourteen faces of the polyhedron. To find such a function exactly is very difficult, but the following approximate method can be used. The polyhedron is nearly a sphere, and we therefore replace it by a sphere of the same volume and use the boundary condition that $\partial u_0/\partial r=0$ at the surface of the sphere. The wave function u_0 is spherically symmetrical in this approximation, and it can be found by integrating the Schrödinger equation numerically if we know the potential produced by the sodium ion at the centre of the polyhedron. The energy determined in this way is shown by the lower curve in fig. 7 as a function of r_s/r_0, where r_s is the radius of the sphere defined above (i.e. $\frac{4}{3}\pi r_s^3$ is the atomic volume), and r_0 is the radius of the first Bohr orbit. The value of a for which the minimum in the energy curve occurs is closely connected with the equilibrium lattice constant, but the two are not identical for the reasons discussed in § 3·2.

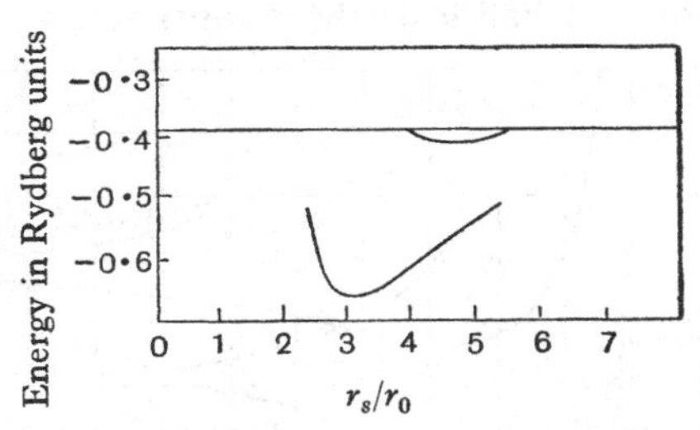

Fig. 7. The energy of the lowest 3s-state (lower curve) and the total energy per atom (upper curve, see p. 34) in sodium as functions of the lattice constant.

The method was extended by Slater (2) to deal with the wave functions and energy levels when $\mathbf{k}=0$. The method consists in solving the Schrödinger equation inside a polyhedron with the proper boundary conditions to ensure that ψ is of the form

$$e^{i(k_1x+k_2y+k_3z)}\, u_{k_1,k_2,k_3}(x, y, z).$$

In practice, however, it is only possible to satisfy the boundary conditions at a finite number of points, usually the centres of the faces of the polyhedron, since otherwise the numerical work becomes prohibitively complicated. By this method we can obtain the energy levels for all $\mathbf{k}$ and a (or r_s). The curves in fig. 8, which is for sodium, give the results. When r_s is infinite the energy levels are discrete, being those of the normal sodium atom, while for any finite value of r_s the typical band structure appears and, for sufficiently small r_s, the bands overlap. (In the diagram only the extreme energy levels corresponding to the top and bottom of each band are shown.) The fact that, for the value of r_s which actually occurs, the first and second bands (the 3s- and 3p-bands) overlap is not of much importance since sodium is monovalent and the valency electrons only half fill the first band. For the alkaline earths, however, the overlapping is not trivial. These metals have two valency electrons per atom, so that they would be insulators if the two lowest bands did not overlap. (This does not apply to Be and Mg since they do not have cubic structures.) It is very satisfactory that calculations by Manning and Krutter (7) for calcium show that there is a decided overlapping of the bands, and therefore that the metallic nature of the alkaline earths is consistent with the theory.

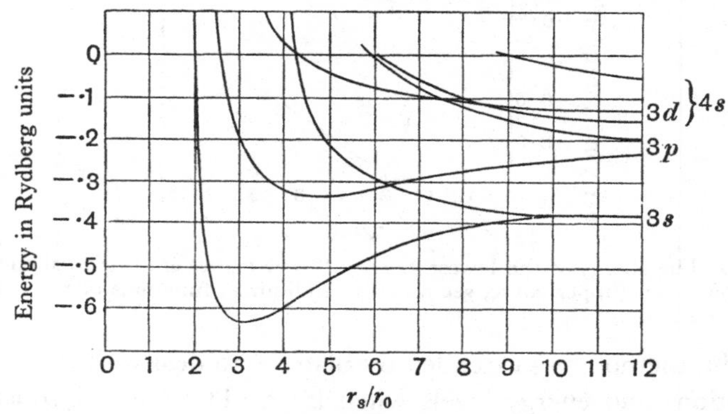

Fig. 8. The energy bands in sodium.

Some of the wave functions calculated by Slater are shown in fig. 9; the thick lines represent the actual wave functions, while the dotted lines represent the wave functions of perfectly free electrons, i.e. $e^{i(k_1x+k_2y+k_3z)}$. It will be seen that over most of the crystal the wave functions are well represented by those for free electrons, particularly for small values of k, the deviations occurring only inside the atomic cores. It is for this reason that many of the properties of metals can be explained so successfully by assuming that the valency electrons behave as if they are perfectly free. A further important point is illustrated by the fact that the curves *b* and *c* have *s*-like properties at some points and *p*-like properties at others.† This means that, when the energy bands overlap, the states can no longer be regarded as being derived from single atomic states; the states have

† A *p* wave function has a node at the centre of an atom while an *s* wave function has not.

become mixed and there is no sharp distinction between the symmetry properties of the wave functions belonging to different bands.

The energy bands have also been found for lithium (3, 4), diamond (5, 6), calcium (7), copper (8, 9), and silver (10); also for LiF and LiH (11) and NaCl (12).

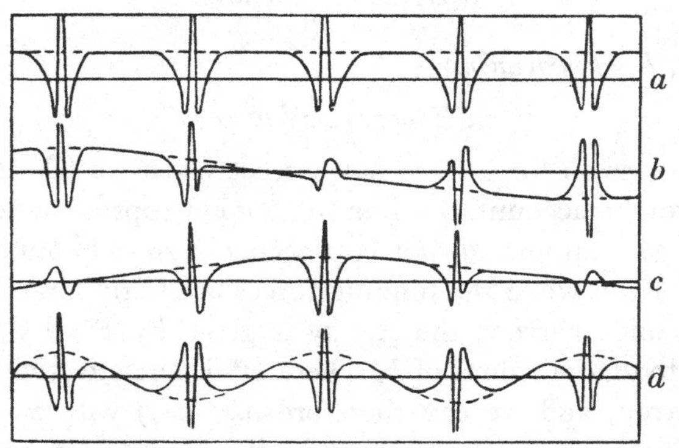

Fig. 9. Wave functions for the 3*s*-electrons in sodium as functions of the distance in the (111) direction. *a*, wave function at the bottom of the band; *b*, *c*, real and imaginary parts of the function whose wave-length is eight times the lattice constant; *d*, wave function at the top of the band.

2·5. The density of states.

Although a knowledge of the energy levels and wave functions of a crystal is essential for a complete description of all its properties, and in particular for the discussion of transport phenomenon, yet knowledge of a much less detailed nature is sufficient for a discussion of the simpler equilibrium phenomena. For example, the specific heat of the electrons can be found if the density of states $\mathfrak{n}(E)$ is known. We define $\mathfrak{n}(E)\,dE$ as being the number of energy levels per unit volume lying in the range E, $E+dE$. Hence

$$\mathfrak{n}(E)\,dE = \frac{1}{8\pi^3}\iiint dk_1\,dk_2\,dk_3, \qquad (2{\cdot}6)$$

the integral being taken over the part of the $\mathbf{k}$ space lying between the energy contours E and $E+dE$. The proportionality factor is fixed by the fact that, when the integral is taken through the fundamental cube $-\pi/a \leqslant k_1, k_2, k_3 \leqslant \pi/a$, the number of energy levels per unit volume must be a^{-3}, the number of atoms per unit volume.

For perfectly free electrons the energy contours are spheres, and, since the volume in **k** space contained between two concentric spheres of radii k and $k=dk$ is $4\pi k^2\,dk$, we have

$$\mathfrak{n}(E)\,dE=\frac{1}{2\pi^2}\,k^2\,dk.$$

Further, $E=h^2k^2/(8\pi^2m)$, so

$$\mathfrak{n}(E)=2\pi(2m)^{\frac{3}{2}}\,h^{-3}E^{\frac{1}{2}}. \qquad (2\cdot7)$$

This expression for $\mathfrak{n}(E)$ is not correct when we take the zone structure into account, but it is still a good approximation when the energy contours do not approach the zone boundaries too closely. Even when the binding forces are fairly large, and the energy bands narrow, the energy is given by $E=h^2k^2/(8\pi^2m^*)$ for not too large values of k, where m^* is the effective mass of the electron, and we can therefore use (2·7) with m^* instead of m. The effect of the binding is, therefore, to increase $\mathfrak{n}(E)$ but to leave the variation with E practically unchanged.

The zone structure has a very marked effect on $\mathfrak{n}(E)$, which is usually exceedingly complicated. Fig. 10 shows $\mathfrak{n}(E)$ when the energy contours are of the type given in fig. 6, p. 20. For small values of E the energy contours are spheres, so that $\mathfrak{n}(E)\propto E^{\frac{1}{2}}$. As the contours deviate from the spherical shape, $\mathfrak{n}(E)$ increases more rapidly than $E^{\frac{1}{2}}$ and reaches a maximum when the contour just touches the zone boundaries. The density of states now rapidly decreases with E until the minimum energy of the second zone is reached; the energy levels of the second zone then start contributing, and $\mathfrak{n}(E)$ increases once more. If the energy zones do not overlap, the density of states in the first zone reaches a maximum and then drops to zero. There follows a region in which there are no energy levels, after which the energy levels of the second zone begin. This behaviour is shown

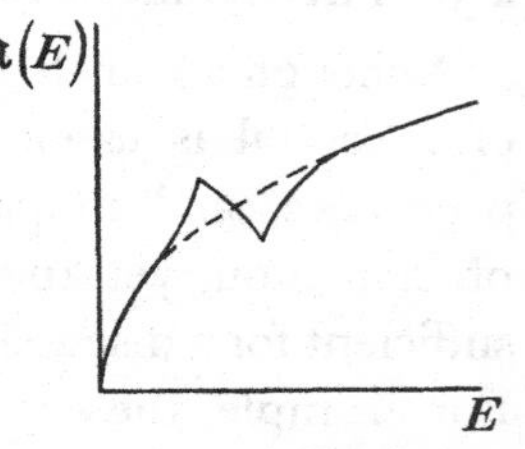

Fig. 10. The influence of the energy discontinuities on the density of states.

in fig. 11 (i). If the two zones only just overlap, as in a semi-metal like bismuth, the form of $\mathfrak{n}(E)$ is as shown in fig. 11 (ii).

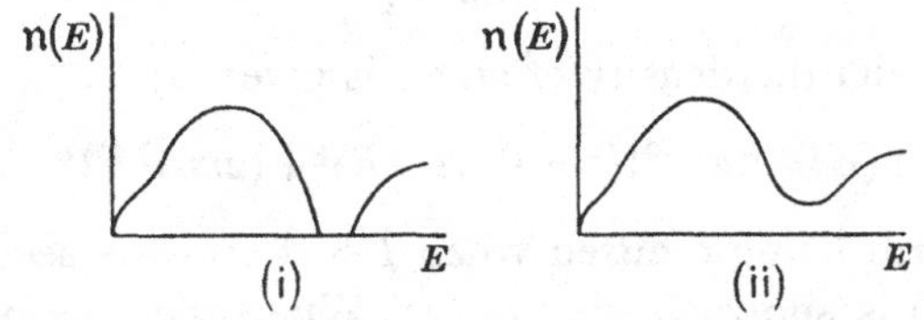

Fig. 11. The density of levels for (i) an insulator and (ii) a semi-metal.

2·51. For convenience in calculation it is often essential to have a simplified model which possesses all the essential features of an actual metal for the problem in hand, but which is such that all the calculations can be carried out exactly and without too much labour. The simplest of these models is that of perfectly free electrons. This is usually a good enough model for mono-valent elements for those phenomena which are the same for all directions in the crystal. This model is quite inadequate for metals in which the electrons partly occupy two bands, in particular for divalent metals. If it is sufficient to assume that

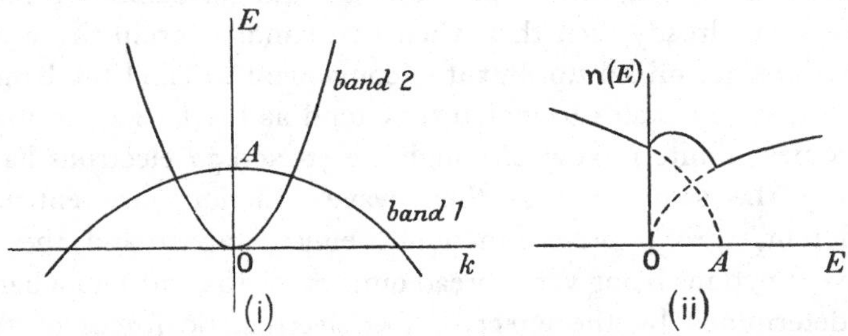

Fig. 12. The energy levels for two overlapping bands of normal form and the density of states.

the metal is isotropic, the simplest model which can be used for a divalent metal is as follows. We take two overlapping bands, as shown in fig. 12 (i), in each of which E is a function of $|\mathbf{k}|$ only. For the lower band, which is nearly filled with electrons, the energy is given by

$$E = A - \frac{h^2 |\mathbf{k}|^2}{8\pi^2 m_1} \quad (A > 0), \tag{2·8}$$

while in the upper band

$$E=\frac{h^2\,|\,\mathbf{k}\,|^2}{8\pi^2 m_2}. \tag{2·9}$$

For this model the density of states is given by

$$\mathfrak{n}(E)=2\pi h^{-3}\{(2m_1)^{\frac{3}{2}}(A-E)^{\frac{1}{2}}+(2m_2)^{\frac{3}{2}}E^{\frac{1}{2}}\}, \tag{2·10}$$

the first term being omitted when $E>A$ and the second when $E>0$; $\mathfrak{n}(E)$ is shown in fig. 12 (ii). When the energy is given by (2·8) and (2·9) we say that the bands are of normal form.

2·52. Of particular interest is the density of states in copper and the neighbouring elements in the periodic table, nickel, cobalt and iron. Copper possesses one 4*s*-electron outside a closed *M* shell, but the existence of divalent copper ions shows that the excitation energy of one of the 3*d*-electrons is at most of the same order of magnitude as the valency forces which exist in chemical compounds, which are also of the same order as the forces in the solid state. This means that, whereas in sodium, for example, the effect of the 2*p*-states on the 3*s*-band can be neglected, in copper the 4*s*- and 3*d*-bands overlap. (We have already seen that when two bands overlap the wave functions get mixed up, but it is convenient to label the bands by the atomic states to which they tend as the lattice constant becomes infinite.) Now, although the 4*s*- and 3*d*-electrons have nearly the same energy, their wave functions are entirely different, the 3*d* wave functions being compact and the 4*s* wave functions being very spread out. Since the width of a band is determined by the effect of the electrostatic forces of the neighbouring atoms on an electron, and since the width therefore depends on the overlapping of the wave functions from one atom to the next, the 4*s*-band should be of approximately the same width as the sodium bands, while the 3*d*-band should be narrow. In other words, the 3*d*-electrons have a large effective mass and move only sluggishly. The density of states in the 3*d*-band is thus abnormally high; in addition there are five 3*d*-states, and this further increases the density of states. Some results of calculations by Slater (13) based on work by

Krutter (8) are shown in fig. 13, $\mathfrak{n}(E)$ being given separately for the 3*d*- and 4*s*-bands. As the figure shows, it is reasonable to assume that the 3*d*-band is of normal form near the top and bottom of the band, but not elsewhere. The numbers 1 to 12 indicate how much of the bands is occupied when there are 1 to 12 electrons per atom. We see that, for copper, the 3*d*-band is full, as we should expect, and that there is one electron per atom in the 4*s*-band. For nickel, on the other hand, not all the

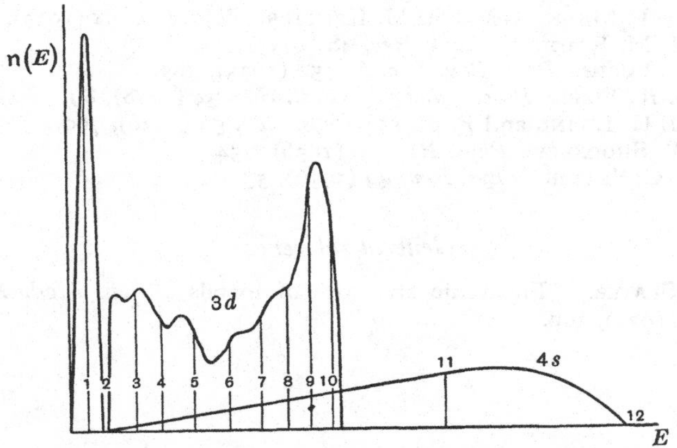

Fig. 13. The density of levels in the 3*d*- and 4*s*-bands.

ten most loosely bound electrons are in the 3*d*-band but some electrons overlap into the 4*s*-band, leaving an equal number of holes in the 3*d*-band. Evidence discussed in § 5·61 indicates that there is about 0·6 electron per atom in the 4*s*-band. That this overlapping should occur is not surprising since in the normal state of the free atom two electrons are in 4*s*-states.

From the above discussion we conclude that the effect of the closed group of 3*d*-electrons will not be noticeable in copper, but that the large density of the 3*d*-states in nickel should have a considerable effect on the properties of the metal.

REFERENCES

(1) E. Wigner and F. Seitz. *Phys. Rev.* **43** (1933), 804 and **46** (1934), 509.
(2) J. C. Slater. *Phys. Rev.* **45** (1934), 794.
(3) J. Millman. *Phys. Rev.* **47** (1935), 286.
(4) F. Seitz. *Phys. Rev.* **47** (1935), 400.
(5) G. E. Kimball. *J. Chem. Phys.* **3** (1935), 560.
(6) F. Hund and B. Mrowka. *Ber. Sächsischen Ges. (Akad.) Wiss. Leipzig*, **87** (1935), 185.
(7) M. F. Manning and H. M. Krutter. *Phys. Rev.* **51** (1937), 761.
(8) H. M. Krutter. *Phys. Rev.* **48** (1935), 664.
(9) K. Fuchs. *Proc. Roy. Soc.* A, **151** (1935), 585.
(10) S. R. Tibbs. *Proc. Cambridge Phil. Soc.* **34** (1938), 89.
(11) D. H. Ewing and F. Seitz. *Phys. Rev.* **50** (1936), 760.
(12) W. Shockley. *Phys. Rev.* **50** (1936), 754.
(13) J. C. Slater. *Phys. Rev.* **49** (1936), 537.

General reference

J. C. Slater. "Electronic structure of metals", *Rev. Mod. Phys.* **6** (1934), 209.

Chapter III

THE STRUCTURE OF METALS

3·1. Cohesive forces in metals.

Solids may, according to the nature of the cohesive forces, be roughly divided into ionic, covalent, molecular and metallic classes, but the division is somewhat arbitrary and there is no such sharp demarcation as exists, for example, between metals and insulators if we take electrical conductivity as the criterion. The cohesion of ionic crystals such as sodium chloride can be readily explained by assuming that, superimposed on the coulomb forces between the ions, there are short-range repulsive forces which prevent the collapse of the crystal. The potential energy of two ions is assumed to be of the form $\lambda r^{-m} - \mu r^{-1}$, where m is a number of the order of 10, the constants λ, μ and m being, in principle, calculable from the crystal structure and the constitution of the ions. The cohesion of molecular crystals is explained in the same way, except that the attractive forces are van der Waals forces instead of coulomb forces, and the potential energy of two molecules is of the form $\lambda r^{-m} - \mu r^{-n}$, where $m > n$ since the forces must be attractive for large r. Examples of molecular lattices are provided by hydrogen and the inert gases. Strong valency forces bind two atoms of hydrogen together to form a molecule which is a saturated structure unable to form bonds with another atom or molecule, and so the only attractive forces between two hydrogen molecules are the weak polarization, or van der Waals, forces.

Diamond affords the simplest example of a covalent structure. Each atom is surrounded by four atoms at the corners of a regular tetrahedron, and, since there are four electrons per atom, each atom can be considered as sharing one electron with each of its four neighbours. The atoms are therefore linked by bonds, so that the whole crystal can be thought of as a gigantic molecule.

In this case the cohesive forces are exchange forces which arise owing to the sharing of two electrons (one from each atom) between two atoms. However, the concept of localized bonds is at best an approximate one and breaks down even in fairly simple molecules, the difficulties in assigning the bonds in the benzene ring being notorious. In benzene three bonds of each carbon atom can be assigned without difficulty, one to a hydrogen atom and two to the neighbouring carbon atoms. This leaves six electrons, one from each carbon atom, to be accounted for. These electrons cannot give rise to bonds of the normal type, and they are to be thought of as mobile electrons shared by the whole of the benzene ring.

The metallic type of binding is the extreme type in which all the most loosely bound electrons are shared between all the atoms in the crystal, there being no connexion between the valency of the substance and the number of nearest neighbours of an atom in the crystal, as in covalent solids. We may therefore say that the binding forces are of infinitely long range, and, since the electrons are shared by all the atoms, we see that the structure of alloys will be determined in the first place by the number of electrons present, and only secondarily by the nature of the atoms from which the electrons come. This is the explanation of the inability of the ordinary valency rules to account for the existence of the so-called intermetallic compounds.

In dealing with questions concerning cohesion it must be kept in mind that all the systems of classification are approximations only. What determines this classification is the nature of the state of lowest energy of the body, and it is only in very simple cases that we can be quite sure what this state is. Even in complex atoms and molecules the energy level system is not easy to unravel, while in crystals it is of enormous complexity. The different categories outlined above are only limiting cases in which the conditions have been so simplified that we can deal with them, and it is unlikely that these conditions apply in their entirety to any actual substances. For example, to obtain covalent binding we must assume that an electron on one atom only interacts with one of the neighbouring atoms, and that these neighbouring atoms do not interact with each other. The metallic binding corresponds to the opposite limiting case in which the effect of distant atoms is as important as that of neighbouring atoms. We must, therefore, not be surprised that many solids refuse to fit into these categories, but appear to have some of the properties both of covalent and of metallic substances.

3·2. The binding energy of sodium.

The methods by which the energy levels of sodium can be obtained have been discussed in § 2·4, and it was shown that the lowest energy level of a free electron has, as a function of the lattice constant, a pronounced minimum. The extra binding energy over that of the free atom is caused by the presence of the atoms surrounding a given atom, which thereby make the wave function periodic with the unit cell as its period, whereas in the free atom the wave function drops rapidly to zero as the distance from the nucleus becomes large. This extra binding energy, however, does not give the total binding energy of the metal, which is the resultant of a number of almost compensating contributions.

The most important repulsive effect comes from the kinetic energy of the electrons, or, in other words, from the fact that all the valency electrons do not occupy the lowest energy level but half fill the 3*s*-band. The magnitude of the kinetic energy is easily calculated if we assume, as is verified by the calculations of Slater on which fig. 8, p. 24, is based, that the effective mass of the electrons is equal to the ordinary mass of an electron. We have in fact for the kinetic energy per unit volume

$$U_{\text{kin}} = 2\int_0^{E_0} E n(E)\,dE, \tag{3·1}$$

where E_0 is the energy of the highest occupied state relative to the bottom of the 3*s*-band. (The factor 2 occurs because each level is occupied by two electrons with opposite spins.) Now, if n is the number of electrons per unit volume, E_0 is determined from the fact that the number of occupied levels per unit volume must be $\frac{1}{2}n$. Hence

$$\tfrac{1}{2}n = \int_0^{E_0} \mathfrak{n}(E)\, dE. \tag{3·2}$$

Now $\mathfrak{n}(E)$ is given by (2·7), so that

$$\tfrac{1}{2}n = 2\pi(2m)^{\frac{3}{2}} h^{-3} \int_0^{E_0} E^{\frac{1}{2}}\, dE = \tfrac{4}{3}\pi(2m)^{\frac{3}{2}} h^{-3} E_0^{\frac{3}{2}}, \tag{3·3}$$

i.e.

$$E_0 = \frac{h^2}{8m}\left(\frac{3n}{\pi}\right)^{\frac{2}{3}}. \tag{3·4}$$

Also

$$U_{\text{kin}} = 4\pi(2m)^{\frac{3}{2}} h^{-3} \int_0^{E_0} E^{\frac{3}{2}}\, dE$$

$$= \frac{3nh^2}{40m}\left(\frac{3n}{\pi}\right)^{\frac{2}{3}}. \tag{3·5}$$

The other contributions are much more difficult to calculate. Since, by the exclusion principle, two electrons with the same spin cannot be in the same state of motion, we see that there is a tendency for electrons with parallel spins to be on the average somewhat far apart. Now our calculations so far have been based on the idea that each electron moves in the average "smeared" field of the ions and of the other electrons, this average being taken without regard to correlations in the positions of the electrons. Since there are in fact such correlations, which tend to make electrons with parallel spins avoid one another, we have overestimated the mutual potential energy of the electrons, which is a repulsion, and to compensate for this we have to introduce an extra binding energy. This energy is called the exchange energy since it is the analogue for free electrons of the exchange energy in molecules. There are, moreover, also correlations between electrons with opposite spins and these introduce a further binding energy which is known as the correlation energy. When all these corrections have been made (1), we obtain for the binding energy per electron as a function of the lattice constant the upper curve shown in fig. 7, p. 23, which has a very flat minimum at $r_s = 2{\cdot}27 \times 10^{-8}$ cm. Since sodium is body-centred cubic and therefore has two electrons in the unit cell, the atomic volume is $\frac{1}{2}a^3$, which gives $r_s = a(3/8\pi)^{\frac{1}{3}}$. Thus the calculated lattice constant is $a = 4{\cdot}62 \times 10^{-8}$ cm. and the binding energy is found to be 1·13 e.volts/atom, i.e. 26 kcal./gram atom. The experimental values are $a = 4{\cdot}24 \times 10^{-8}$ cm. and 30 kcal./gram atom for the binding energy. Such close agreement is almost certainly fortuitous, and the results for other metals are not so good.

3·3. The compressibility.

The compressibility κ is determined by the curvature of the energy curve at the minimum, and, as fig. 7 shows, the minimum occurs at a point where the energy of the lowest state is nearly a linear function of a. The curvature is actually mainly determined by the variation of U_{kin}, and we can obtain the order of magnitude of κ by neglecting the other parts of U. Since the pressure p is given by $p = -d(U/n)/dv$, where v is the atomic volume and n is the number of atoms per unit volume (thus U/n is the internal energy per atom), we have

$$\frac{1}{\kappa} \equiv -v\frac{dp}{dv} = v\frac{d^2}{dv^2}\left(\frac{U}{n}\right). \tag{3·6}$$

Now $v = \frac{4}{3}\pi r_s^3$, so

$$\frac{1}{\kappa} = \frac{r_s}{3}\frac{d}{dr_s}\left\{\frac{1}{4\pi r_s^2}\frac{d}{dr_s}\left(\frac{U}{n}\right)\right\} = \frac{1}{12\pi r_s}\frac{d^2}{dr_s^2}\left(\frac{U}{n}\right), \tag{3·7}$$

since $d(U/n)/dr_s = 0$ is the condition for equilibrium.

If we now substitute for U the expression (3·5) for U_{kin}, and remember that $n^{-1} = v = \frac{4}{3}\pi r^3$, we find that

$$\frac{1}{\kappa} = \left(\frac{9}{4\pi}\right)^{\frac{2}{3}} \frac{3h^2}{80\pi m r_s^5} = \frac{3^{\frac{2}{3}} h^2}{10 m a^5}. \tag{3·8}$$

The calculated and observed values of κ for the alkalis are given in Table I; the observed values are for temperatures ranging from 30° to 50° C. The agreement is as good as can be expected considering the rough approximations used. Another result which is obvious from fig. 7 is that for high pressures the compressibility is determined largely by the energy curve for the lowest state, which has a large curvature near the minimum.

Thus the compressibility of the alkalis must have a very large (negative) pressure-coefficient.

TABLE I. Compressibilities of the alkalis in (kg./cm.^{2})$^{-1}$

	Li	Na	K	Rb	Cs
a in 10^{-8} cm.	3·46	4·24	5·25	5·62	6·05
$\kappa \times 10^6$ observed	8·7	15·6	35·7	52	70
$\kappa \times 10^6$ calculated	5	14	40	56	81

The compressibility of the noble metals is low, κ for copper being $7{\cdot}2 \times 10^{-7}$ (kg./cm.^{2})$^{-1}$. This is because, owing to the small lattice constant, the atomic cores are almost in contact, whereas in the alkalis the structure is more diffuse and the cores are very far apart. The expression (3·8) is, therefore, not even approximately correct for the noble metals; their compressibility is determined almost entirely by the deformability of the cores.

3·4. Survey of the metallic structures.

Although we possess considerable knowledge of the energy levels of several metals, it is not yet possible to predict the crystal structure of a metal, the reason being that the differences in energy between the various structures are less than the errors in the numerical calculations which have so far been carried out. It is clear that for sufficiently short-range forces the face-centred cubic (i.e. cubic close packed) structure is more stable than the body-centred cubic, since the number of nearest neighbours, the coordination number, is twelve for the former and only eight for the latter. Since, however, the metallic forces are long-range forces the paucity of nearest neighbours in the body-centred lattice is compensated by the much greater number of second nearest neighbours as compared with the face-centred lattice. The approximate method of Wigner and Seitz gives, in fact, the same energy for both structures. For, their method consists essentially in solving the Schrödinger equation inside a sphere whose volume is the atomic volume, and hence the energy value calculated by them depends only on the atomic volume and not on the crystal structure.

The only properties which vary widely from metal to metal

are those which depend upon the magnitude of the energy discontinuities at the zone boundaries and upon the relation of the highest occupied energy levels to these boundaries. Now it is just for these energy levels that our quantitative knowledge is weakest, and so all that we can do at present is to interpret the properties of metals in terms of the distribution of the energy levels, not to predict them.

3·41. Typical metallic structures.

The majority of the elements crystallize in one of the metallic structures, the body-centred cubic, the face-centred cubic and the hexagonal close packed, the cells of which are shown in fig. 14. The coordination numbers of the lattices are as follows.

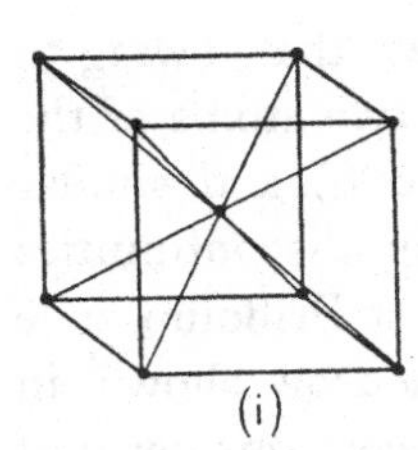
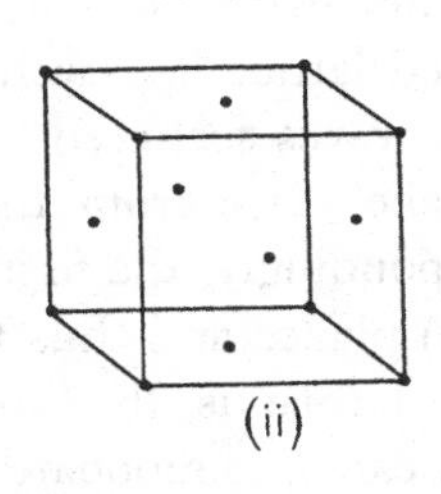
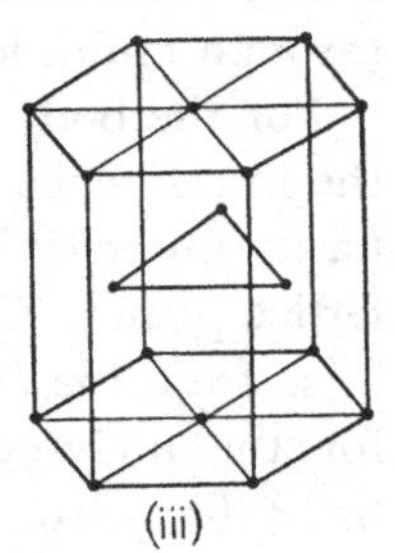

Fig. 14. The structures of (i) body-centred cubic, (ii) face-centred cubic and (iii) hexagonal close packed lattices.

In the body-centred structure each atom has eight nearest neighbours at a distance $\frac{1}{2}\sqrt{3}\,a$, the atoms at the centres of the cubes being the neighbours of those at the corners. In the face-centred structure the atom at (0, 0, 0) has as its nearest neighbours the twelve atoms at the face centres such as $(\pm\frac{1}{2}a, \pm\frac{1}{2}a, 0)$, which are at a distance $\sqrt{2}\,a$. In the hexagonal structure each atom has six neighbours at a distance a, which lie in the basal plane through the atom perpendicular to the hexagonal axis. There are six others at a distance $\sqrt{(\frac{1}{3}a^2+\frac{1}{4}c^2)}$, three lying above and three below the basal plane. If the axial ratio $c/a=(8/3)^{\frac{1}{2}}$, these twelve neighbours are all at the same distance, the case of ideal close packing; the coordination number is then twelve, while, if c/a deviates very much from the ideal ratio, the coordination number is six.

It is clear that the binding in these lattices cannot be covalent and must be metallic in character. Thus, a body-centred structure, for example, could only be covalent if each atom shared one of its electrons with each of its eight neighbours, which would require far too much energy for the structure to be stable.

3·42. The simplest way of dealing with the body- and face-centred cubic lattices is to regard them as lattices with a basis. For the former the basis consists of two atoms at (0, 0, 0) and $(\frac{1}{2}a, \frac{1}{2}a, \frac{1}{2}a)$, while for the latter the basis consists of four atoms at (0, 0, 0), $(\frac{1}{2}a, \frac{1}{2}a, 0)$, $(\frac{1}{2}a, 0, \frac{1}{2}a)$ and $(0, \frac{1}{2}a, \frac{1}{2}a)$. The whole lattice is obtained from the basis by translations along the axes through multiples of the lattice constant.

For the body-centred lattice it is well known that, owing to the interference of the waves scattered by the two atoms of the basis, no reflexions take place from the (1, 0, 0) and similar lattice planes. Correspondingly, the first energy discontinuities arise from the (1, 1, 0) planes and thus the first Brillouin zone for the body-centred lattice is the dodecahedron shown in fig. 5 (ii), p. 19, which can accommodate four electrons per unit cell of the crystal. There are two atoms per unit cell, and hence in the alkalis the valency electrons just fill half of this zone. Now we should expect the energy contours to be approximately spherical unless they approach the zone boundaries closely. If we assume them to be spheres, the highest occupied level has a radius k_0, where $\frac{4}{3}\pi k_0^3 : (2\pi/a)^3 = 2 : 2$, since the fundamental cube in $\mathbf{k}$ space of side $2\pi/a$ can accommodate two electrons per atom and the number of energy levels in a region of $\mathbf{k}$ space is proportional to its volume. Hence $k_0 = (6\pi^2)^{\frac{1}{3}}/a = 3{\cdot}898/a$. The radius of the sphere inscribed in the dodecahedron is equal to the distance of the planes (2·3) from the origin, i.e. it is $\sqrt{2}\pi/a = 4{\cdot}44/a$, which is considerably larger than k_0. Thus the approximation of treating the energy contours of the occupied levels as being spheres, i.e. of treating the electrons as free electrons, is seen to be a reasonable one for the alkalis.

Little can be said about the energy contours of the other

body-centred metals, except that the zone structure must reduce the metallic character of the substances. The valency electrons of barium, for example, are just sufficient to fill up the first zone, but in fact they do not do so. Instead they overlap into the second zone, and the energy discontinuities must be fairly small.

In face-centred structures the (1, 0, 0) and the (1, 1, 0) reflexions do not occur. Hence the boundaries of the first zone are related to the (1, 1, 1) and (2, 0, 0) reflexions, the zone being the truncated octahedron shown in fig. 5 (iv), p. 19, which can accommodate eight electrons per unit cell. Since there are now four atoms per unit cell, in the noble metals the valency electrons just fill half this zone. Just as for the alkalis, it is a reasonable approximation to assume that the energy contours are spheres, and that the valency electrons occupy a sphere of radius $k_0 = (12\pi^2)^{\frac{1}{3}}/a = 4\cdot91/a$. The radius of the sphere inscribed in the Brillouin zone is equal to the distance of the planes (2·4) from the origin, i.e. it is $\sqrt{3}\,\pi/a = 5\cdot44/a$. The energy contour of the highest occupied level therefore approaches the zone boundary more closely in the face-centred than in the body-centred structure, and the approximation of free electrons is presumably not so good.

The first Brillouin zone of the hexagonal close-packed lattice is somewhat complicated. It consists of a hexagonal prism surmounted by two truncated hexagonal pyramids and is shown in fig. 15. Unlike the zones of the cubic lattices, it does not contain an integral number of energy levels since its volume depends on the axial ratio. By actual mensuration it can be shown that this zone can accommodate

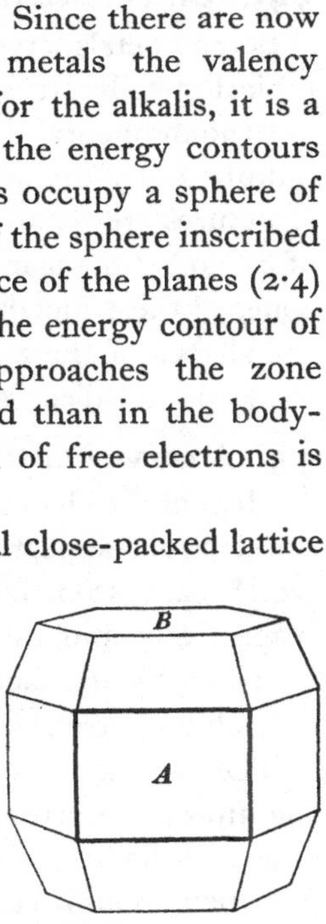

$$2 - \frac{3}{4}\left(\frac{a}{c}\right)^2 \left\{1 - \frac{1}{4}\left(\frac{a}{c}\right)^2\right\} \qquad (3\cdot9)$$

electrons per atom. For the axial ratios which occur, (3·9) is less than 2, and so the electrons of all the divalent hexagonal

Fig. 15. The first Brillouin zone of a hexagonal close-packed lattice.

metals overlap into the second zone. To take an actual example, only 1·792 electrons per atom can be accommodated in the first zone of zinc, the axial ratio being 1·856. Further, since the zone is fairly symmetrical, we should expect that the number actually in the first zone would not be very much less than 1·79, so that there may be only something of the order of 0·3 electron per atom in the second zone. Thus, although zinc has two valency electrons, we should expect it to be much less metallic in character than the alkalis.

3·43. Other metallic structures.

Some metals crystallize in slightly distorted forms of the typical metallic structures. Examples are mercury, whose lattice is a rhombohedral distortion of a face-centred cubic lattice, and indium and γ-manganese which have tetragonal lattices. Other structures are those of white tin and the complicated structures of α- and β-manganese. It is not difficult to find the Brillouin zones of these metals, but it has not proved possible to correlate the shapes of the zones with the properties of the metals, and so we do not discuss them here.

3·5. A survey of the structures of the semi-metals.

Most of the elements with incomplete groups of four, five and six electrons appear at first sight to form covalent structures in which each atom has four, three and two nearest neighbours respectively, but that this is not an exact description of affairs is shown by the fact that many of these elements have a considerable electrical conductivity.

The elements selenium and tellurium have lattices in which the atoms are arranged in spiral chains as shown in fig. 16, so that each atom has two close neighbours in each chain, and four more distant neighbours in adjacent chains. In tellurium the interatomic distances are $2{\cdot}86 \times 10^{-8}$ cm. and $3{\cdot}46 \times 10^{-8}$ cm. Thus if we are to describe the lattice as covalent, the forces in each chain must be covalent while the chains are kept together by van der Waals forces.

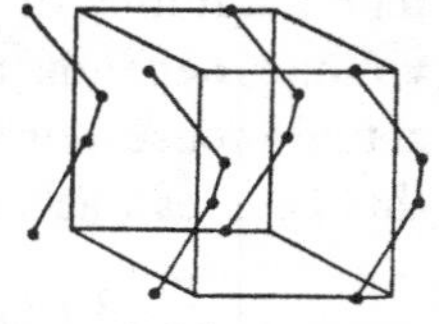

Fig. 16. The structure of selenium.

The structures of metallic phosphorus, arsenic, antimony and bismuth are rhombohedral in which the atoms are arranged in double layers. Each atom has three near neighbours in the double layer in which it is situated and three others at a greater distance in an adjacent layer. If the forces keeping each double layer together are covalent forces, the forces between the layers must be van der Waals forces, but since the interatomic distances—between nearest and next nearest neighbours—are $1{\cdot}74 \times 10^{-8}$ cm. and $3{\cdot}01 \times 10^{-8}$ cm. in phosphorus, while they are nearly equal ($3{\cdot}105 \times 10^{-8}$ cm. and $3{\cdot}474 \times 10^{-8}$ cm.) in bismuth, the description of the structure as covalent may be a reasonable approximation for the former substance but it must be a bad approximation for the latter.

Although the semi-metals can be classed neither as covalent nor as metallic substances, yet, if we have to choose between the two descriptions, the metallic one seems to be the better, though quantitative calculations have not yet been carried out to confirm this. If the metallic description of these substances is to be at all adequate, we should expect there to be a Brillouin zone just capable of holding all or nearly all the electrons. As we have seen, this is a question which is settled by the crystal structure alone, and it is not difficult to find Brillouin zones which are capable of holding five electrons per atom in the case of the bismuth structure and six in the case of the selenium structure. In addition to this, it is necessary, if these solids are to have properties intermediate between those of metals and insulators, that the energy discontinuities should be neither too small nor too large. No quantitative calculations of the discontinuities have yet been carried out, and all that we can do is to infer from the known properties of the substances that the discontinuities are such that only very few electrons overlap into the higher zones, leaving a corresponding number of holes in the almost full zone. Evidence discussed in § 6·25 suggests that the number of electrons in the second zone in bismuth is about 10^{-3} per atom.

The remaining semi-metals are carbon, silicon, germanium and (grey) tin, which have diamond lattices. For this structure there is a zone which has twice the linear dimensions of the dodecahedron shown in fig. 5 (ii), p. 19, and which can just accommodate four electrons per atom. The energy discontinuities at the surface of this zone must be fairly large in diamond and must decrease with increasing atomic number, since diamond is the only one of these substances which is an insulator. Graphite, the other form of carbon, has a complicated sheet-like structure, the atoms in each sheet forming regular hexagons. Thus graphite bears the same relation to the benzene structures as diamond does to the ordinary tetrahedral carbon compounds. It behaves as a highly anisotropic semi-metal, which is what we should expect, since the forces between the atoms in a sheet must be very different from the forces holding the sheets together.

3·6. Alloys.

When metals are melted together they may or may not mix, and, if they do, it does not follow that a homogeneous solid solution will be obtained on cooling. The possible behaviour of even binary alloys is so complex that the factors which influence the formation and stability of alloys have only been elucidated in some very simple cases.

One of the important factors in alloy formation is the relative size of the metal atoms, since, if the atoms differ too much in size, it is impossible to arrange them regularly so as to form an alloy devoid of cavities. Another important factor is the "electron concentration". It has been found that certain types of crystal structure tend to be formed when the ratio of the number of valency electrons to the number of atoms reaches certain fairly precise values. For example, two empirical rules enunciated by Hume-Rothery are that the body-centred cubic β-brass structure tends to be formed when the electron concentration (the value of the ratio above) is $\frac{3}{2}$, and that the complex cubic γ-brass structure tends to be formed when the electron concentration is $\frac{21}{13}$. It has already been pointed out in § 3·1 that the cohesive forces in metals and also in alloys are to be thought of as belonging to the whole system of valency electrons and not as being localized between neighbouring atoms. It is, therefore, clear that, since the energy levels of the electrons depend upon a large number of factors, we should expect few regularities in the behaviour of alloys. For those alloys which obey the Hume-Rothery rules, however, it has been found that the most stable structure at any concentration is that which possesses a Brillouin zone capable of accommodating the valency electrons and leaving the largest number of levels vacant. For further details concerning these alloys the reader is referred to any of the more recent text-books on metals, but it must be emphasized that the Hume-Rothery rules are by no means of universal application and that the theory of alloys is still only in a primitive state.

REFERENCES

(1) E. Wigner. *Phys. Rev.* **46** (1934), 1002.

General references

W. Hume-Rothery. *The metallic state.* (Oxford, 1931.)

W. Hume-Rothery. *The structure of metals and alloys.* (London, 1936.)

F. C. Nix and W. Shockley. "Order-disorder transformations in alloys", *Rev. Mod. Phys.* **10** (1938), 1.

Chapter IV

SEMI-CONDUCTORS

4·1. General principles.

In § 1·3 we saw that an insulator is a substance in which the electrons just fill an energy band at the absolute zero. At any higher temperature some of the electrons are thermally excited into the next, unoccupied, band, leaving some vacant spaces in the occupied band. The number of electrons excited must be of the form $e^{-b/T}$, where b is connected with the energy step between the two bands, and since these electrons are free to move through the lattice they produce an electrical conductivity whose temperature variation is also of the form $e^{-b/T}$. Insulators such as sodium chloride do show a weak electrical conductivity, but this is due to the electrolytic conduction of positive ions, as is shown by the polarization effects set up by the passage of the current. There are, however, substances, conventionally known as semi-conductors, which have at room temperatures an appreciable electronic conductivity which varies rapidly with temperature. These substances, which are of great technical importance, owe their properties to the presence of impurities (this was first clearly recognized by Gudden), and therefore it is not possible to deal theoretically with all the details of their behaviour, but the general principles are well understood.

Many substances such as graphite, silicon, titanium and zirconium were formerly thought to be semi-conductors since their resistances decreased as the temperature increased. This anomalous behaviour is now known to be due to the presence of highly insulating layers of oxides, and these substances are good metals or at worst semi-metals in the pure state. The only reliable test for distinguishing between a semi-conductor and a metal is that the conductivity of a metal is increased by purification, while the conductivity of a semi-conductor is reduced by the removal of impurities.

The ways in which impurities can affect the conductivity of insulators can be most readily understood with the help of fig. 17, in which two energy bands of a crystal are shown. The band F is completely filled with electrons when the temperature is zero, and the band E is completely empty. Consider the effect of an impurity atom D which has one valency electron, the energy level of this electron lying between the two bands. This electron cannot take part directly in conduction if the number of impurities is small, since to do so it would have to jump to a similar state on another impurity atom. The probability of such a jump decreases very rapidly as the distance between the atoms increases, and is negligible for the concentrations with which we are concerned. The electron on the impurity atom can only take part in conduction by being first thermally excited into the empty band E, when it is free to move through the lattice. In this case the conductivity still varies as $e^{-b/T}$, but b is now connected with the energy difference between the level D and the band E and not with the energy difference between the two bands. Thus by introducing suitable impurities we may be able to make b so small that the substance has an appreciable electronic conductivity at room temperatures. We distinguish four important cases.

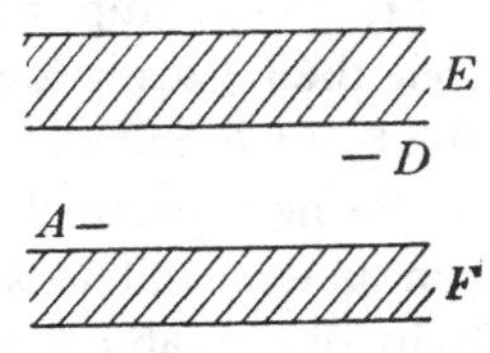

Fig. 17. The energy levels in a semi-conductor.

(1) If the impurity atoms are electropositive and the energy levels of their valency electrons lie between the two bands, the atoms act as *donors* of electrons to band E, and the number of electrons in band E is of the form $e^{-b/T}$.

(2) If the impurities are electropositive and the energy levels lie in the band E, the atoms are permanently ionized and the number of electrons in band E is constant, in so far as the electrons are derived from the impurities.

(3) If the impurity atoms are electronegative they may have vacant energy levels, such as A in fig. 17, lying between the two

bands. In this case the impurity atoms can act as *acceptors* of electrons from the band F; holes are created in band F and their number is proportional to $e^{-b/T}$, where b is connected with the energy difference between the band F and the energy level A.

(4) If the impurity atoms are electronegative and are such that their vacant levels lie in the band F, holes are created in the band F and their number is constant.

We now proceed to discuss some of the properties of semi-conductors and to see how far they can be understood in the light of the above principles. It must, however, be borne in mind that there is considerable dispute about many of the experimental details.

4·2. Criteria for establishing the nature of the conductivity.

Since both the electronic and electrolytic conductivities in semi-conductors are proportional to $e^{-b/T}$, we cannot decide the nature of the conductivity by measurement of the temperature coefficient of the conductivity alone; other evidence must be invoked. Polarization effects and departures from Ohm's law are evidences of electrolytic conductivity and their absence is conclusive proof of the existence of electronic conductivity. However, even when there appears to be a transport of positive ions obeying Faraday's law, it cannot necessarily be concluded that the conductivity is overwhelmingly electrolytic. This is shown by the dispute concerning the behaviour of α-Ag_2S, which is the modification stable above 179° C. That the conductivity is electrolytic seemed to be established by the fact that the amount of silver removed from a silver anode in contact with the sulphide is exactly that given by Faraday's law. On the other hand, the conductivity does not vary rapidly with temperature and is about fifty times greater than that of any known electrolytic conductor. Further, the mobility of the silver ions calculated from the conductivity is very much greater than that calculated from diffusion measurements. The discrepancy was finally explained away by Wagner(1), who showed that the disappearance of silver from the anode was a purely secondary diffusion phenomenon not directly connected with the passage of the current through the body of the sulphide. It now seems certain that about 99 per cent of the current in α-Ag_2S is carried by electrons.

For substances in which the conductivity has been proved to be electronic, it is very important to know whether the current is carried by electrons in a nearly empty band or by "holes" in a band nearly filled with electrons. Since the "holes" behave in many ways like positive electrons, this question can be settled by measuring any quantity which depends on the first power of the charge of the carrier; such quantities are the Hall coefficient and the thermoelectric power (§§ 6·3, 6·5). It must, however, be remembered that the current may in some substances be carried both by electrons and by "holes". In this case measurements of the Hall effect merely show which is the more important method of transport. If we call conductors in which the current is mainly carried by electrons in a nearly empty band "excess conductors", and those in which it is carried mainly by holes "defect conductors", the following are some of the substances in the two classes (2–11). Excess conductors: α-Ag_2S, MoS_2, ZnO, Al_2O_3, Ta_2O_5, V_2O_5, WO_3. Defect conductors: Cu_2O, CuI, UO_2, NiO.

4·3. Oxidation and reduction semi-conductors.

Many of the most important semi-conductors are oxides and, even in their purest forms, are slightly dissociated; they owe their properties to the presence of a small excess of oxygen or of the metal above the stoichiometric amount. These substances can be classified according to the way in which they behave when the oxygen content is varied.

Substances whose conductivity increases with the oxygen content are called oxidation semi-conductors, since the conductivity is due to the presence of excess oxygen, though it is not perfectly clear what is the exact mechanism of producing the conductivity. The oxygen atoms may act directly as acceptors of electrons or, in say cuprous oxide, they may first abstract electrons from cuprous ions so as to turn them into doubly charged cupric ions, which then act as acceptors. Whatever the mechanism, the ultimate unit which is responsible for the conductivity is electronegative, and oxidation semi-conductors should be defect conductors in which the conductivity is due to holes in a nearly filled band.

Reduction semi-conductors are those in which the conductivity decreases as the oxygen content increases. Their conductivity is therefore due to the presence of excess metal atoms which act as donors of electrons.

The few experiments which have been carried out confirm the conjecture that oxidation conductors are defect conductors and that reduction conductors are excess conductors (but not necessarily the converse). A further rule has been formulated by Friederich (12), Le Blanc and Sachse (13), Meyer (14) and Wagner (15), which states that compounds in which the metal exerts its smallest valency are oxidation conductors, while saturated compounds in which the metal exerts its highest valency are reduction conductors. This rule has been proved to be correct for the defect conductors Cu_2O, CuI, NiO and UO_2 in which the acceptor atoms are probably divalent copper, trivalent nickel and sextivalent uranium, and also for the excess conductors α-Ag_2S, ZnO, Al_2O_3 and Ta_2O_5. It is probably true in other cases such as CdO, FeO, CoO, TiO_2, but complete measurements of the Hall effect (or the thermoelectric power) and of the influence of the composition on the conductivity have not yet been carried out.

The above rule does not appear to be true in all cases, and there are some substances such as cupric oxide whose conductivity is independent of the oxygen content. The significance of the exceptions is at present obscure.

4·4. The dependence of the conductivity on the amount of impurity.

The conductivity of most semi-conductors can be well represented by the formula $\sigma = ae^{-b/T}$ or $\sigma = a' T^s e^{-b/T}$ over a large temperature range. The two formulae are indistinguishable experimentally, and, though the second is to be preferred on theoretical grounds, the index s cannot be determined except by making certain assumptions whose validity is open to question. Measurements of σ alone do not give the number of free electrons since σ is given by $\sigma = n\epsilon^2\tau/m$, where τ is the mean time of relaxation (see (6·2)). However, only n varies exponentially with temperature, and so the temperature variations of n and σ are practically the same. If great accuracy in b is desired, it is better to use measurements of the Hall coefficient R, since, by (6·10), we have

$$R = -\frac{3\pi}{8n\epsilon c} \qquad (4\cdot1)$$

for excess conductors, while for defect conductors the minus sign is to be omitted. For conductors in which the current is carried partly by electrons and partly by holes, the numbers of the carriers cannot be determined with any accuracy.

It seems reasonable to suppose that the quantity a should be determined by the number of impurity atoms present, while the quantity b should be independent of that number and should depend only on the nature of the impurity, since b is connected with the energy difference between the levels of the impurities and of the bulk material. This reasoning is based on the assumption that cuprous oxide, say, in which there are excess oxygen atoms, behaves, for small concentrations, like an ideal dilute solution. Now the experimental results show that the energy kb varies within wide limits, which can only mean that the above assumption is wrong. For cuprous oxide, kb has been found by various workers to lie between 0·1 and 0·4 e.volt, the great majority of the measurements being near 0·3 e.volt. The reasons why b is not constant are not at present understood, and this makes a complete quantitative theoretical treatment impossible. Two possible explanations suggest themselves. First, the additional oxygen atoms may be able to exist only where there are cracks or faults in the material. They would there be subject to distorting forces which would alter the binding energy of the electrons, and therefore also b, in an irregular manner. Secondly, the oxygen atoms may be arranged in clusters, so that nuclei of cupric oxide are formed. In this case b might be expected to vary smoothly with the oxygen content, but experimental evidence to test this point is lacking.

4·41. The action of the impurities in producing free electrons can be described by the equation

$$\text{free electron} + \text{positive ion} \rightleftarrows \text{bound electron} \qquad (4\cdot 2)$$

for excess conductors, and by the equation

$$\text{free hole} + \text{negative ion} \rightleftarrows \text{bound hole} \qquad (4\cdot 3)$$

for defect conductors. (In these equations positive ion and negative ion are to be taken in a relative sense. Thus Cu^+ may have to be regarded as the negative ion of Cu^{++}.) Since excess conductors are easier to think of, we restrict ourselves to the first case, but the results are the same for both. Let N be the number of impurity atoms per unit volume, and n the number of free electrons. Then if n_+ is the number of positive ions with which the free electrons can combine, the law of mass action gives

$$\frac{nn_+}{N-n} = K, \qquad (4\cdot 4)$$

where K is the equilibrium constant, since the number of bound electrons per unit volume to which the right-hand side of (4·2) refers is $N-n$. The difficulty in proceeding further is that we do not know what n_+ is. There are, however, two limiting cases which are easily discussed. (1) If the free electrons can only recombine with the impurities from which they came, then $n_+ = n$. Since K, apart from powers of T which are unimportant, varies as $e^{-\Delta E/kT}$, where ΔE is the energy of activation, i.e. the energy difference between the impurity level and the empty band of the bulk material, we have

$$n \propto N^{\frac{1}{2}} e^{-\frac{1}{2}\Delta E/kT}, \qquad (4\cdot 5)$$

provided that $n \ll N$. (2) If the free electrons can combine with any of the positive ions in the crystal, then n_+ is practically equal to the number of metal ions present and can therefore be treated as a constant. We then have

$$n \propto N e^{-\Delta E/kT}. \qquad (4\cdot 6)$$

These are the extreme cases. If the free electrons can combine with only some of the metal ions, say those situated at cracks, as well as the original impurities, then we have an intermediate case.

4·42. The two points raised by the theory, namely the dependence of n on N and on ΔE, are not easily tested experimentally. In the first place the amount of impurity present can only be found directly when it is sufficiently large to be detected chemically, and in the second place the amount of impurity is not necessarily the same as the number of available electrons if the impurities occur in large clusters. (Only the atoms on the outside of the cluster would be effective.) The only experiments which have been carried out to test the theory directly are those of Wagner (5, 6, 8) and Bädecker (16). Wagner measured the conductivity of cuprous oxide in equilibrium with an oxygen atmosphere at temperatures between 800° and 1000° C. He found that the conductivity is a single-valued function of the oxygen pressure and that approximately

$$\sigma \propto (p_O)^{1/7}. \tag{4·7}$$

The chemical reaction can be represented by the equations

$$O_2 + 4Cu^+ \rightleftarrows 2O^{--} + 4Cu^{++}, \tag{4·8}$$

$$4Cu^{++} \rightleftarrows 4Cu^+ + 4 \text{ free holes}. \tag{4·9}$$

The equation (4·9) is the same as (4·3); the cupric ions abstract electrons from the full band of electrons and produce holes which are free to carry a current. If square brackets are used to denote concentrations, we have from (4·8) that

$$[Cu^{++}] \propto (p_{O_2})^{1/4}. \tag{4·10}$$

In order to determine n, the number of free holes per unit volume, we have to make some assumption concerning the number of Cu^+ ions which can combine with the free holes. If we make the assumption which leads to (4·5), then $[Cu^+] = n$, and so

$$n \propto (p_{O_2})^{1/8}. \tag{4·11}$$

This is in fair agreement with (4·7) and is an indication that the reaction can be considered as a homogeneous reaction between holes and impurities. This is confirmed by the fact that Wagner found that kb for the conductivity at constant oxygen content had the value 0·4 e.volt and was independent of the oxygen content. It is in any case clear that the experimental results cannot be reconciled with the assumption which leads to (4·6), namely that the number of cuprous ions which can combine with the free holes is very much larger than the number of free holes, since we should then have $\sigma \propto (p_{O_2})^{1/4}$.

Similar experiments have been carried out for the reduction conductor ZnO. Since the conductivity is caused by excess zinc atoms, the conductivity is reduced by increasing the oxygen pressure, and it is found that between 400° and 700° C.

$$\sigma \propto (p_{O_2})^{-1/4\cdot3} \qquad (4\cdot12)$$

approximately. The chemical equations are probably

$$O_2 + 2Zn \rightleftharpoons 2ZnO, \qquad (4\cdot13)$$

and

$$Zn \rightleftharpoons Zn^{++} + 2 \text{ free electrons.} \qquad (4\cdot14)$$

Hence the law of mass action gives

$$p_{O_2}[Zn^{++}]^2 n^4 = K, \qquad (4\cdot15)$$

and, if we assume that $n = [Zn^{++}]$, we have

$$n \propto (p_{O_2})^{-1/6}. \qquad (4\cdot16)$$

This does not agree at all well with the experimental result (4·12). In fact better agreement is obtained if we assume that $[Zn^{++}]$ is independent of n, or, what is more plausible, that the zinc is not completely ionised.

4·5. The photoelectric effect.

Most of the work on semi-conductors has been carried out, not in equilibrium with oxygen gas, but at fairly low temperatures (less than a few hundred degrees C.) with a constant amount of oxygen "frozen in". In this case the inner photoelectric effect has been used to give information concerning the number of impurity atoms and the energy levels.

The absorption of light by cuprous oxide has been thoroughly investigated by Schönwald (17) and Engelhard (18). It was found that in addition to the absorption band with a maximum at $0\cdot63\mu$, which is characteristic of the pure substance, there are weak absorption bands with maxima at $0\cdot8\mu$ and 2μ. The significance of the band at 0·8μ is not known, but the band at 2μ is definitely associated with the ordinary conductivity observed in the dark. At first sight these measurements appear to give us a means of determining ΔE, the energy required to create a free hole, and of comparing it with the value of kb found from the electrical conductivity. There are, however, certain difficulties in this procedure. In the first place the infra-red absorption band is of considerable width, and we have to decide whether to take the long wave-length limit or the position of the maximum as determining ΔE. The long wave-length limit lies at about 4μ, so it is a large energy difference which is in question, and the matter can really only be decided if we know exactly what the energy levels are. Moreover, it is known that for substances like the alkali halides the infra-red absorption bands become narrower as the temperature is decreased, and the maximum shifts to higher frequencies. The second difficulty is discussed below.

In the two extreme cases considered in § 4·41 we should have $kb=\frac{1}{2}\Delta E$ according to (4·5), and $kb=\Delta E$ according to (4·6). Now kb is about 0·3 e.volt and the wave-length 2μ corresponds to an energy of 0·6 e.volt while 4μ corresponds to 0·3 e.volt. Thus either of the two alternative hypotheses can be made to fit in with the facts,† and neither can be definitely proved or disproved. For other substances neither hypothesis agrees well with the facts. It was pointed out by de Boer and van Geel (19) that in general we should expect no simple relation to exist between the thermal activation energy and the photoelectric threshold energy $h\nu_0$. The first is an energy difference between two equilibrium states while, on account of the Franck-Condon principle, the second is not. When an electron is removed thermally (i.e. slowly) from an atom the crystal becomes distorted near that atom, but when the electron is excited by light the lattice does not have time to take up this distorted form, and so the two energies are not the

† The results would prove that $kb=\frac{1}{2}\Delta E$ if the excitation energy is determined by the position of the maximum of the absorption band, and that $kb=\Delta E$ if it is the long wave-length limit which is important.

same. It therefore appears that direct comparison between thermal and photoelectric data is not always possible.

The dependence of the photoconductivity on the intensity of light is more easy to interpret. It is found that the extra conductivity induced, and therefore the number of electrons excited, is proportional to the intensity I. Now if N is the number of impurities, the rate at which electrons are excited is proportional to NI. The recombination rate lies between those given by the two limiting cases of § 4·41, namely constant $\times n^2$ and constant $\times n$. Thus in a steady state the extremes of the dependence of n upon I should be given by (1) $n \propto N^{\frac{1}{2}} I^{\frac{1}{2}}$ and (2) $n \propto NI$. The experiments show that in fact n is proportional to I, and thus that, if the same ions are responsible for recombination in the thermal as in the photoelectric case (and there is no reason to suppose otherwise), (4·6) is to be preferred to (4·5). This is to be contrasted with the conclusion, based on Wagner's results, that (4·5), though not in exact agreement with the experiments, is to be preferred to (4·6) at high temperatures.

4·6. Other possible tests of the theory.

There are two classes of semi-conductors for which many of the difficulties of interpretation are considerably lessened. The first consists of semi-conductors of the type (2) described in § 4·1, i.e. semi-conductors in which the number of free electrons is independent of the temperature and equal to the number of impurities. It is, however, very doubtful whether such substances exist, the only substance so far investigated (4) which might belong to this class being α-Ag_2S. The rhombic β-modification, which is stable below 179° C., behaves like a normal semi-conductor in which some of the current is carried electrolytically. The conductivity of the cubic α-modification is about 300 times larger than that of β-Ag_2S just below the transition point, and it decreases slowly and linearly as the temperature increases. This decrease is presumably due to the decrease in the free path, as in metals, the number o free electrons remaining constant. The properties of silver sulphide have not been sufficiently investigated for us to be able to say whether this interpretation is correct or not, and it is possible that here we are dealing with a semi-conductor of the normal type, but in which ΔE is small compared with kT, so that, at the temperatures at which the measurements are carried out, all the impurities are ionized. Since α-Ag_2S is only stable at high temperatures, this possibility cannot be tested by measurements at low temperatures.

The other simple type of semi-conductor is that in which the free electrons are derived from the atoms of the pure substance and not from impurities. It has been claimed by Jusé and Kurtschatow (20) that extremely pure cuprous oxide comes under this category. The inner photoelectric effect shows that to raise an electron from the filled band to the empty band in cuprous oxide requires about 2 e.volts. Since this is so much larger than the 0·3 e.volt (or 0·6 e.volt according to (4·5)) required to transfer an electron to a surplus oxygen atom, the intrinsic conductivity could only be observed for a very pure specimen, in which the intrinsic conductivity would not be masked by the "impurity conductivity", and at a high temperature. There is, however, one factor which favours the intrinsic conductivity. The electrons from the filled band can be excited from any of the doubly charged oxygen ions of the crystal, whereas only the impurities are effective in the ordinary case and the number of impurities is only of the order 10^{18} to 10^{20} per $cm.^3$ Therefore if we write $\sigma = ae^{-b/T}$, both a and b are much larger for the intrinsic conductivity than for the "impurity conductivity", and the largeness of b is partly compensated by the largeness of a. Jusé and Kurtschatow claim to have measured σ for very pure cuprous oxide and also for cuprous oxide containing a small amount of oxygen at temperatures high enough for the intrinsic conductivity to be dominant. They found that $kb = 0{\cdot}72$ e.volt. The number of measurements was, however, small, and their results have not been confirmed by other experimenters. At the moment it is an open question whether the intrinsic conductivity is observable or not.

If it should prove possible to observe the intrinsic conductivity, we could test the general assumptions of the theory without having to make any special hypotheses. When electrons are excited from the filled to the empty band the number of holes is equal to the number of electrons and hence (4·5) ought to apply, except that N is to be taken as the total number of oxygen atoms and not the number of excess atoms. In this case, therefore, we have $kb = \frac{1}{2}\Delta E$. Deviations from the behaviour predicted by the simple theory would indicate the formation of such things as metastable excited atoms and neutral copper atoms, but such speculations belong to the future.

4·61. Other methods of investigation do not so much test the theory of semi-conductors as test the theory of the effect measured. This is certainly true of the change of resistance in a magnetic field (see § 6·34) and to a less extent of the thermoelectric power. The thermoelectric force per degree $d\Theta_{12}/dT$ for a couple formed of two specimens of different purity is given by (*T.M.* p. 182, equation (267))

$$\frac{d\Theta_{12}}{dT} = \mp \frac{k}{\epsilon} \log \frac{n_2}{n_1}, \qquad (4\cdot 17)$$

the minus sign referring to excess conductors and the plus sign to defect conductors.

The measurements of Wagner (6) are in reasonable agreement with this formula, the measured thermoelectric force for two specimens of cuprous oxide being 3×10^{-5} volt/degree at 900° C. and the calculated value being $3{\cdot}6 \times 10^{-5}$ volt/degree.

The thermoelectric force of semi-conductors is large in comparison with that of metals (see § 6·53), and hence the absolute thermoelectric force of a semi-conductor can be found fairly accurately by using a couple consisting of a metal and the semi-conductor, and not correcting for the contribution of the metal. The most important part of the thermoelectric force of a semi-conductor comes from the term $\log 1/n$ (compare (4·17)), and since n is of the form $n_0 e^{-b/T}$, we have

$$\frac{d\Theta}{dT} = \text{constant} \mp \frac{kb}{\epsilon T}. \qquad (4\cdot 18)$$

The first term is not really a constant, but its variation with T is small, and in any case it is small compared with the second term.

Hochberg and Sominski [11] have recently carried out measurements on the temperature dependence of $d\Theta/dT$ for a number of substances. They find that (4·18) holds for WO_3 and also, but not so well, for V_2O_5. For MoS_2 and Cu_2O, on the other hand, they find that $d\Theta/dT$ is practically constant. It is well known that the theory of the thermoelectric effects is not very satisfactory even for metals, and the lack of agreement between (4·18) and the experimental results may be due to some neglected factor in the theory. However, the experimental accuracy is not great, and the results do not agree with those of Vogt [21] for cuprous oxide. Vogt found that the thermoelectric power of cuprous oxide against copper was about 1·3 mV./degree at room temperature and that it increased considerably as the temperature decreased. The temperature variation between $-60°$ and $70°$ C. was compatible with the $1/T$ law, but the results were not sufficiently precise either to confirm or refute it with certainty.

4·7. Crystal rectifiers.

One of the most important technical applications of semi-conductors is to the rectification of alternating currents. It has been known since 1874 that, when a metal point is placed in contact with a semi-conductor, either in the form of a crystal or as tarnish on wires, a current passes through the circuit more easily in one direction than the other.

The best known commercial rectifier, the cuprous oxide rectifier discovered by Grondahl in 1920, owes its advantages to the method of manufacture. (For a full description of the rectifier see [22].) A sheet of copper is partially oxidized at a high temperature, so that a layer of cuprous oxide is formed which has a very intimate contact with the mother copper. Thus rectification takes place over a large area, and the rectifier is capable of carrying large currents. By measuring the potential difference between various parts of the rectifier, it has been found that there is a highly insulating layer, the barrier layer, between the oxide and the metal, and that the presence of this layer is essential for the occurrence of rectification. Since the conductivity of cuprous oxide is due to the presence of excess oxygen, it is highly probable that the barrier layer consists of very pure, and therefore highly insulating, cuprous oxide. It would indeed be difficult for an excess of oxygen to exist in the immediate neighbourhood of the mother copper.

Since the resistance of a homogeneous substance is necessarily the same in one direction as in the opposite, the asymmetry in the resistance of the rectifier can be due only to the "contact resistance". A theory of the contact resistance which, until recently, seemed to be most promising, is based on the idea that, provided that the barrier layer is not too thick, it can be treated as a potential hump through which electrons can pass by means of the tunnel effect. We give an outline of this theory and then discuss the difficulties which beset it.

Although cuprous oxide is really a defect conductor, it is so much simpler to think of an excess conductor that we consider the semi-conductor to be of the latter type. The only levels with which we are concerned in the semi-conductor are those in the first incompletely filled band, there being in fact a few free electrons in this band which have a Maxwell distribution. The lowest energy levels in the metal and semi-conductor, and the potential hump, are shown in fig. 18. In the metal there is a very large number of electrons which have a Fermi distribution, while in the semi-conductor there are very few electrons, the distribution being Maxwellian. In equilibrium there must be no resultant current in either direction, and so the surfaces charge up until the number of electrons leaving the metal per second is the same as that leaving the semi-conductor. Now, by the principle of detailed balancing, this can happen only when the number entering each individual energy level is the same as that leaving, and hence equilibrium is reached when the energy levels are displaced so that the Maxwell distribution of the semi-conductor just corresponds to the tail of the Fermi distribution. (Since no electrons can enter the forbidden levels of the semi-conductor, there is no question of having to balance the electrons in the metal whose levels lie lower than the lowest allowed and unoccupied level in the semi-conductor.) When the metal is made negative with respect to the semi-conductor, the energy of the electrons in the former is increased and hence the Fermi distribution is raised relatively to the Maxwell distribution. There is therefore a resultant electron current from the metal to the semi-conductor, some of the electrons going over the potential hump, but many passing through the hump (the well-known tunnel effect). Similarly when the potential is reversed

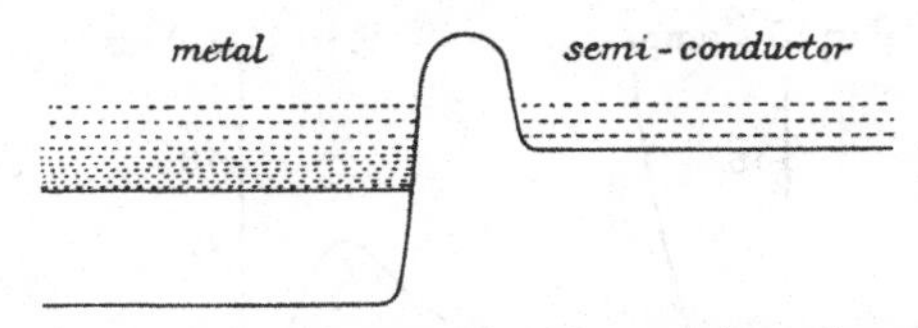

Fig. 18. The energy levels of a metal and a semi-conductor in contact in equilibrium. The density of the dots indicates roughly the number of electrons thermally excited.

there is a resultant current from the semi-conductor across the junction; but, whereas effectively an unlimited number of electrons can leave the metal, the same is not true of the semi-conductor since there are so few electrons present. In other words, giving the metal a negative potential brings into play a number of electrons which previously could not enter the semi-conductor since the energy levels which they occupied in the metal are forbidden in the semi-conductor. Thus it is possible for a very large current to pass from the metal to the semi-conductor, while in the reverse direction the current is necessarily small, since there are very few electrons in the semi-conductor and the number available does not depend on the field.

Some curves calculated for a special type of hump are shown in fig. 19, the second curve referring to the differential resistance $R = dV/dJ$, where V is the potential drop from the metal to the semi-conductor and J is the current per unit area. It will be noticed that R has a maximum for negative V, i.e. when the current is from the semi-conductor across the gap. The reason for this is as follows. The penetrability of the potential hump decreases as the energy of the electron decreases, and so the slow electrons in the semi-conductor do not have much chance of getting into the metal. Now it can be shown that the effect of a strong electric field is to increase the transparency of the barrier and thus the slower electrons have a better chance of getting out of the semi-conductor in this case; the resistance therefore decreases. If the transmission coefficient of the barrier were unity this effect could not, of course, occur.

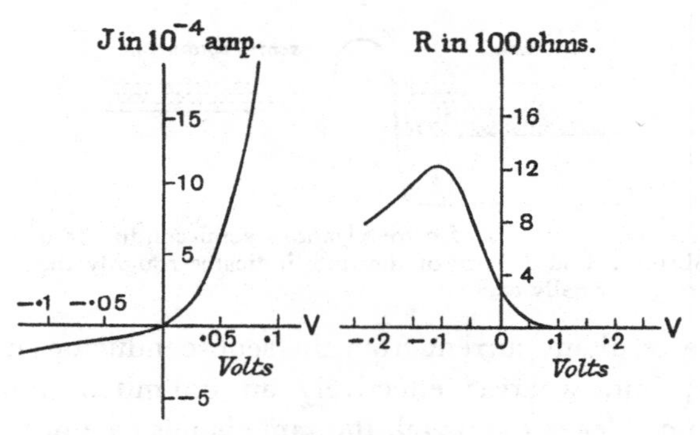

Fig. 19. The current and differential resistance as a function of the voltage for a contact between a metal and a semi-conductor.

At first sight it might appear that the rectification which takes place at one contact between the metal and semi-conductor would be nullified at the second contact. (We necessarily have to consider a closed circuit.) This would be true if both contacts were identical, but rectifiers are made so as to have one good contact (no potential hump) and one bad contact (a large potential hump), and it is the latter at which the rectification takes place. When a current is flowing, the potential drop has to distribute itself over the circuit so as to produce the same current at all points. Thus there must be a large potential drop at the bad contact and a much smaller one at the good contact, and since, as we see from fig. 19, there is very little asymmetry in the current for small voltages, it is only the bad contact which is important in the rectification.

In order to make the above discussion apply to cuprous oxide we must replace "electrons" by "holes"; the easy direction of flow of holes is still from the metal to the semi-conductor, since there is an unlimited number of holes available in the metal. This gives rise to a very serious difficulty since the observed easy direction of flow of holes is from the oxide to the metal. (The above theory was put forward at a time when the experiments were supposed to give a negative Hall coefficient and thus to indicate that cuprous oxide was an excess conductor. The theory was in fact devised so as to make the direction of easy flow from the metal.) Another difficulty is that recent measurements put the thickness of the barrier layer as high as 10^{-3} cm. This means that the transparency of the potential hump is quite negligible and that the electrons can only pass over the hump. The theory given above only applies when the width of the hump is of the order 10^{-7} or 10^{-6} cm.; thicknesses of the latter order of magnitude were indicated by the early measurements.

Perhaps the most promising attempt to modify the theory so as to give the right direction of rectification is that due to Schottky (23), but as yet only an abstract has been published. The theory seems to be on the following lines. Since the barrier is so wide that electrons can only go over the top of the potential hump, the shape of the hump is important. The arrangement of the energy levels is as shown in fig. 18, except that the hump is wide. Also the hump probably slopes gradually down to the conduction levels of the oxide from a maximum value situated at or near the metal surface. If now the levels in the oxide are raised by applying an external field, the potential hump is diminished and the number of electrons which can flow over the hump into the metal is increased. On the other hand, when the levels in the metal are raised the number of electrons which can flow from the metal is unchanged. Thus the direction of easy flow is from the oxide to the metal. According to this theory, therefore, the rectification is caused by the differential effect of the applied field in modifying the potential hump, and it is essential that the maximum of the hump should occur very near the metal surface. Further details are necessary before the theory can be properly appraised.

4·8. The discussion of semi-conductors given in this chapter, though by no means complete, shows that the general theoretical ideas have considerable qualitative value. On the other hand, the behaviour of semi-conductors is so individual and varied that only experiment can decide which of the almost infinite number of possibilities is realized in any particular case.

REFERENCES

(1) C. WAGNER. *Zeit. phys. Chem.* B, **21** (1933), 42.
(2) K. STEINBERG. *Ann. Physik* (4), **35** (1911), 1009
(3) C. W. HEAPS. *Phil. Mag.* (7), **6** (1928), 1283.
(4) F. KLAIBER. *Ann. Physik* (5), **3** (1929), 229.
(5) H. H. v. BAUMBACH and C. WAGNER. *Zeit. phys. Chem.* B, **22** (1933), 199.
(6) H. DÜNWALD and C. WAGNER. *Zeit. phys. Chem.* B, **22** (1933), 212.
(7) H. H. v. BAUMBACH and C. WAGNER. *Zeit. phys. Chem.* B, **24** (1934), 59.
(8) K. NAGEL and C. WAGNER. *Zeit. phys. Chem.* B, **25** (1934), 71.
(9) O. FRITSCH. *Ann. Physik* (5), **22** (1935), 375.
(10) W. HARTMANN. *Zeit. Phys.* **102** (1936), 709.
(11) B. M. HOCHBERG and M. S. SOMINSKI. *Phys. Z. Sowjet,* **13** (1938), 198.
(12) E. FRIEDERICH. *Zeit. Phys.* **31** (1925), 813.
(13) M. LE BLANC and H. SACHSE. *Phys. Z.* **32** (1931), 887.
(14) W. MEYER. *Zeit. Phys.* **85** (1933), 278.
(15) C. WAGNER. *Zeit. phys. Chem.* B, **22** (1933), 181.
(16) K. BÄDECKER. *Ann. Physik* (4), **29** (1908), 566.
(17) B. SCHÖNWALD. *Ann. Physik* (5), **15** (1932), 395.
(18) E. ENGELHARD. *Ann. Physik* (5), **17** (1933), 501.
(19) J. H. DE BOER and W. C. VAN GEEL. *Physica* (2), **2** (1935), 286.
(20) W. JUSÉ and B. W. KURTSCHATOW. *Phys. Z. Sowjet,* **2** (1932), 453.
(21) W. VOGT. *Ann. Physik* (5), **7** (1930), 183.
(22) L. O. GRONDAHL. *Rev. Mod. Phys.* **5** (1933), 141.
(23) W. SCHOTTKY. *Naturwissenschaften,* **26** (1938), 843. Also N. F. MOTT, unpublished.

General reference

B. GUDDEN. "Elektrische Leitfähigkeit elektronische Halbleiter", *Ergebnisse der exakten Naturwissenschaften* 13. (Berlin, 1934.)

Chapter V

THE THERMAL AND MAGNETIC PROPERTIES OF METALS

5·1. The specific heat of the electrons.

We saw in § 1·42 that only those electrons whose energy levels lie in an interval of the order kT round E_0, where E_0 is the Fermi energy, contribute to the specific heat of a degenerate electron gas. If $\mathfrak{n}(E)$ is the density of states, the number of electrons per unit volume in the range kT is $kT\mathfrak{n}(E_0)$, and since each electron contributes $\frac{3}{2}k$ to the specific heat, the total specific heat of the electron gas per unit volume is of the order

$$k^2T\mathfrak{n}(E_0). \tag{5·1}$$

An accurate calculation gives (*T.M.* p. 185, equation (268))

$$C_v = \tfrac{2}{3}\pi^2k^2T\mathfrak{n}(E_0), \tag{5·2}$$

and if we assume that the electrons behave as if they were free, with an effective mass m^*, $\mathfrak{n}(E)$ is given by (2·7) and (1·13), so that

$$C_v = \frac{4\pi^3m^*k^2}{3h^2}\left(\frac{3n}{\pi}\right)^{\frac{1}{3}}T, \tag{5·3}$$

where n is the number of free electrons per unit volume. In general the specific heat of the electrons is small compared with that of the lattice vibrations, but, since the contribution of the latter is proportional to T^3 at low temperatures, the electronic specific heat must be the more important for sufficiently small T.

It is only recently that any measurements have been made of the specific heat of the electrons [1–4]. The results are shown in Table II, expressed in calories/degree/gram atom. In order to compare the theoretical and experimental results we have to change the units in (5·2) and (5·3) by dividing by n_a, the number of atoms per unit volume, and multiplying by Loschmidt's number L. We obtain for the specific heat per gram atom

$$C_v' = \tfrac{2}{3}\pi^2 k\mathfrak{n}(E_0)\, n_a^{-1} RT$$
$$= 17{\cdot}9 \times 10^{-16}\mathfrak{n}(E_0)\, n_a^{-1} T \text{ cal./deg./g. atom,} \qquad (5{\cdot}4)$$
$$= \frac{4\pi^3 m^* k}{3h^2 n_a}\left(\frac{3n}{\pi}\right)^{\frac{1}{3}} RT = 23{\cdot}7 \times 10^{10}\, \frac{m^* n^{\frac{1}{3}}}{m n_a} T \text{ cal./deg./g. atom.} \qquad (5{\cdot}5)$$

The values of $\mathfrak{n}(E_0)/n_a$ and of m^*/m deduced from the experimental results are given in Table II. Since $\mathfrak{n}(E_0)/n_a$ is given in absolute units by (5·4) we have to multiply the numbers

TABLE II. Specific heats of metals at low temperatures in units of $10^{-4}T$ cal./deg./g. atom

	C_v'	$n_a \times 10^{-22}$	$\mathfrak{n}(E_0)/n_a$ in (electron volts)$^{-1}$	m^*/m
Cu	1·78	8·5	0·16	1·5
Ag	1·6	5·9	0·14	1·0
Zn	(2·5)	6·7	(0·22)	(1·4)
Hg	3·7	4·3	0·33	1·5
Al	3·5	6·1	0·3	1·6
Tl	3·8	3·6	0·33	1·2
Sn	3·5	4·3	0·3	1·1
Pb	7·1	3·3	0·6	1·9
Ta	27	5·7	2·4	—
Nb	(70)	5·6	(6)	—
Ni	17·4	9·2	3·1	35
Pd	31	6·9	2·7	27
Pt	16·1	6·7	1·4	14

Notes. The values of C_v' for Hg, Tl, Sn, Pb, Ta, Nb are lower limits, since they are obtained from the entropy differences between the superconductive and non-superconductive states by assuming that $C_v'=0$ for the superconductive state. Since nickel is ferromagnetic, only the states with one direction of spin contribute to C_v' at low temperatures (see §§ 5·2 and 5·6). The value of $\mathfrak{n}(E_0)$ obtained from (5·4) must therefore be doubled.

obtained by $1{\cdot}59 \times 10^{-12}$ in order to reduce them to the number of levels per atom per electron volt. In compiling this table we have taken n to be equal to the number of valency electrons per unit volume except for the transition elements; for nickel we take $n = 0{\cdot}6n_a$ and for palladium and platinum $n = 0{\cdot}5n_a$ (see §§ 5·52 and 5·61). For all the metals in Table II except Cu and Ag, the electrons occupy more than one zone, and therefore m^* has not very much meaning, since it is derived from a formula which only holds if the electrons are perfectly free, i.e. if the zone structure is unimportant. Further, the d-bands in say nickel (see below) may be degenerate. In this case the calculated value of m^* contains as a factor the weight of the band, which may be as much as 5. On the other hand $\mathfrak{n}(E_0)$ is a well-defined quantity and the best way of determining it is by means of the specific heat at low temperatures.

The large values of $\mathfrak{n}(E_0)/n_a$ for the transition metals can only be due to the large density of states in the d-band. We saw in § 2·52 that in nickel there is a wide $4s$-band with a small $\mathfrak{n}(E)$ and a narrow $3d$-band with a large $\mathfrak{n}(E)$ which overlaps the $4s$-band, and that the number of electrons is such that there are some electrons in the s-band and some vacant spaces in the d-band. A similar situation occurs in the other transition elements, and in all cases the large density of states is due to a narrow d-band. An estimate of $\mathfrak{n}(E_0)$ for nickel has been given by Slater (5) based on calculations by Krutter, and he finds that $C_v' = 0{\cdot}0011\,T$, which is somewhat less than, but of the same order of magnitude as, that found experimentally.

5·2. The temperature variation of the specific heat.

The proportionality of C_v to T is only true provided that $T \ll T_0$, where T_0 is the degeneracy temperature, and at very high temperatures C_v must tend asymptotically to the classical value of $\frac{3}{2}nk$. The exact form of C_v as a function of T can only be found by numerical calculations, the most complete of which are due to Stoner (6), but, since some assumption must be made concerning the behaviour of $\mathfrak{n}(E)$ as a function of E, the results of these calculations have not the same validity as the low temperature result (5·2). The usual assumption is that the energy band is of normal form, so that we have $\mathfrak{n}(E) \propto E^{\frac{1}{2}}$, or, if we are dealing with a nearly full band, we have $\mathfrak{n}(E) \propto (A-E)^{\frac{1}{2}}$, where A is the energy at the top of the band. The results are shown in fig. 20, p. 67, where C_v is given as a function of T/T_0.

The only temperatures where we can hope to measure the

electronic specific heat accurately are either very low or very high ones; the first because the specific heat of the lattice is negligible there, and the second because the Debye specific heat is constant at high temperatures. There are, however, two difficulties in the interpretation of the high temperature results. These are (1) that it is C_p that is measured and not C_v, and (2) that the Debye theory is only an approximation and the specific heat of the lattice vibrations does not really become constant. The correction to constant volume can be estimated semi-empirically, but there is at present no reliable way of estimating what effect the anharmonic coupling of the lattice vibrations has on the specific heat, and this difficulty cannot be overcome. For example, the measurements of Jaeger, Rosenbohm and Bottema (7) indicate that in copper the excess specific heat over the Debye value is 0·37 cal./degree/gram atom at 1000° K. If this is an electronic specific heat, it is much too large to be reconcilable with the low-temperature value (Table II), and it therefore indicates either that the present theory is inadequate, or that the effect of the anharmonic coupling is fairly large.

5·21. The specific heat of nickel.

Nickel is the only metal for which detailed measurements have been carried out, but the interpretation of the results is somewhat difficult because nickel is ferromagnetic. To offset this, the electronic specific heat is large, and so the effect of the anharmonic coupling of the lattice vibrations is relatively less important than it appears to be in copper. A detailed analysis of the specific heat of nickel has been carried out by Stoner (8) and compared with the theoretical predictions. Since nickel is ferromagnetic it has an excess specific heat over the Debye value, due to the destruction of the intrinsic magnetization by the thermal agitation; this specific heat increases up to the Curie point and then drops to zero. There is still a fairly large excess specific heat above the Curie point and this is interpreted as being due to the electrons. Of course the "ferromagnetic" specific heat is also an electronic specific heat, but it is due to the change in direction of the spins of the electrons, and, for convenience, we restrict the term "electronic specific heat" to mean a specific heat not associated with a change in spin. It is, however, not entirely clear that this separation of the specific heat into a ferromagnetic or spin contribution and a translational contribution is justified.

Since the electronic configuration is different above and below the Curie point, any comparison of the high- and low-temperature specific heats involves some assumption about this configuration. This is touched upon in § 5·61; the assumption usually made is that only states with one direction of spin contribute to the specific heat at low temperatures, while both sets of states contribute above the Curie point. Thus the density of states

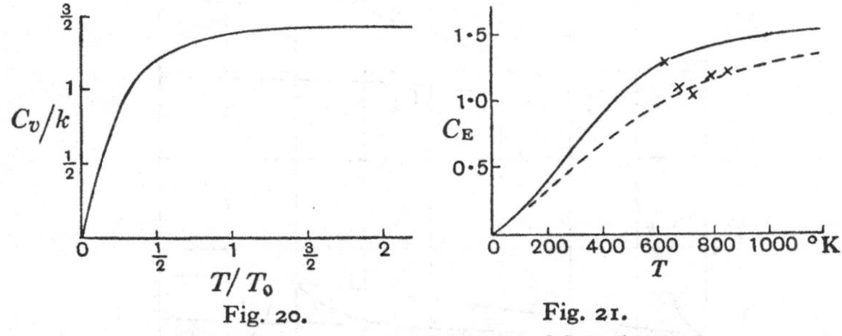

Fig. 20. Fig. 21.

Fig. 20. The specific heat per electron of a gas of free electrons.

Fig. 21. The electronic specific heat of nickel in cal./degree/gram atom. The continuous curve is calculated on the assumption that the magnetism varies, and the broken curve is calculated on the assumption that the magnetization does not vary. The experimental values are shown by crosses.

above the Curie point is larger than the low temperature values and the specific heat is correspondingly increased. Further, since the density of states is so large, the electron gas can no longer be treated as completely degenerate at high temperatures, T_0 being about 2000° K. The electronic specific heat as calculated by Stoner is shown in fig. 21. The calculations are based on the low-temperature value $1{\cdot}74\times10^{-4}T$ for the specific heat, and the numbers of the electrons with opposite spins at any temperature are chosen so as to give the observed magnetization at that temperature.

The experimental values, which are due to Lapp (9), Grew (10), Klinkardt (11) and Sykes and Wilkinson (12), and which are not corrected for the anharmonic contribution of the lattice, are in general agreement with the calculations, but the position is not entirely satisfactory. According to the most recent measurements (12), better agreement is obtained if we calculate the specific heat above the Curie point on the assumption that the number of electrons with a given direction of spin is independent of the temperature (see fig. 21); this assumption is, of course, untenable. The source of the discrepancy may be that it is incorrect to split up the specific heat into a magnetic part and a trans-

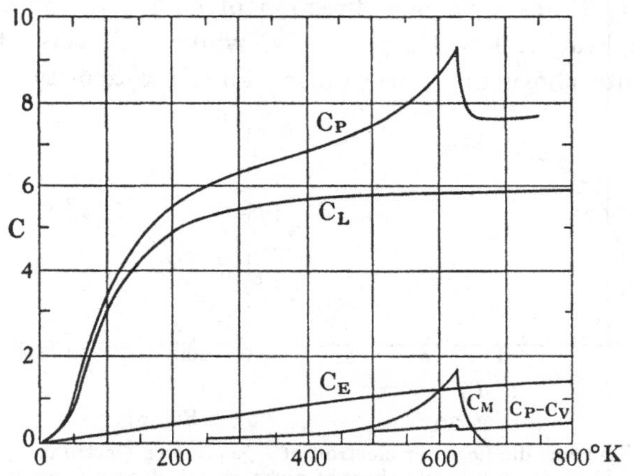

Fig. 22. The specific heat of nickel in cal./degree/gram atom. C_P is the observed specific heat; C_L is the lattice specific heat; C_E is the electronic and C_M is the magnetic specific heat. The last three are at constant volume.

lational part, or it may be that the degeneracy temperature has been wrongly estimated by assuming that the bands are of normal form. A further complication is that if the anharmonic contribution of the lattice has the same value as in copper (0·2 cal./degree/gram atom), then the experimental values of the electronic specific heat must be still further reduced. There seems to be no method of estimating the anharmonic contribution theoretically, and, to make matters worse, the contribution can be negative as well as positive. At the present time, therefore, we are still far from being able to say that the theory has been verified quantitatively. The various contributions to the specific heat as calculated by Stoner are shown in fig. 22.

5·3. The spin paramagnetism.

Langevin's classical theory of the paramagnetism of a substance composed of molecules having a permanent magnetic moment μ_0 leads to the following formula for the susceptibility χ per unit volume:

$$\chi = \frac{N\mu_0^2}{3kT}, \tag{5·6}$$

where N is the number of molecules per unit volume. This formula fits well with the experimental results for gases and for salts of the rare earths, but it is not even approximately correct for metals. In general, χ for metals is small and nearly independent of the temperature, whereas, since μ_0 for an electron is one Bohr magneton $\epsilon h/4\pi mc$, formula (5·6) gives much too large a susceptibility at ordinary temperatures. This difficulty is removed when we apply Fermi-Dirac statistics instead of classical statistics to the free electrons in a metal.

The effect of a magnetic field H is to increase the number of electrons with their spins along the direction of the field at the expense of those with their spins in the opposite direction, the transfer of an electron from one spin state to the other lowering the potential energy by $2\mu_0 H$. This transfer, however, increases the kinetic energy, and the alignment of the electrons only proceeds until the diminution of the potential energy is equal to the gain in kinetic energy. Since the electrons whose spins have been reversed must necessarily go into states which were previously unoccupied, it is clear that only those electrons near the top of the Fermi distribution can contribute to the paramagnetism, and that the electrons in question are those whose energies lie in a range of the order $2\mu_0 H$ round E_0. This energy interval is independent of T, and so χ in first approximation is constant. Since each electron contributes μ_0 to the magnetic moment, and since the number of the energy levels concerned is $2\mu_0 H\mathfrak{n}(E_0)$ per unit volume, the total magnetic moment per unit volume is $2\mu_0^2 H\mathfrak{n}(E_0)$, and hence

$$\chi = 2\mu_0^2\mathfrak{n}(E_0). \tag{5·7}$$

For free electrons we have, by (1·13) and (2·7),

$$\chi = \frac{3n\mu_0^2}{2E_0}. \tag{5·8}$$

Both C_v and χ at low temperatures are proportional to $\mathfrak{n}(E_0)$, and there is therefore a relation between the electronic specific heat and the spin paramagnetism:

$$\frac{C_v}{\chi} = \frac{\pi^2}{3}\left(\frac{k}{\mu_0}\right)^2 T, \quad (T \ll T_0). \tag{5·9}$$

5·31. The temperature variation of the spin paramagnetism.

When the temperature is so high that the classical statistics is applicable we must have

$$\chi = \frac{n\mu_0^2}{kT}, \quad (T \gg T_0). \tag{5·10}$$

This differs from Langevin's formula (5·6) by the absence of the factor $\frac{1}{3}$. This factor arises as the mean value of $\cos^2\theta$, where θ is the angle between the magnetic moment of the molecule and the direction of the field. Classically all orientations are possible

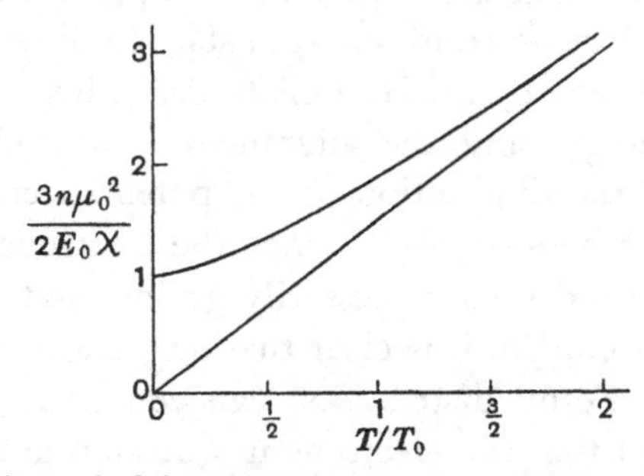

Fig. 23. The reciprocal of the paramagnetic susceptibility as a function of the temperature for a gas of free electrons. The straight line gives the classical value of $1/\chi$.

and the mean value of $\cos^2\theta$ is correctly taken to be $\frac{1}{3}$, but for electrons only two orientations are possible, namely $\theta = 0$ or π, and the mean value of $\cos^2\theta$ is 1.

The exact way in which χ passes over from (5·7) to (5·10) depends upon the behaviour of $\mathfrak{n}(E)$ as a function of E. Numerical calculations have been carried out by Stoner (13) for the case in which the energy band is of normal form so that $\mathfrak{n}(E) \propto E^{\frac{1}{2}}$, and his results are shown in fig. 23. He has also given the second approximation to χ in the form

$$\chi = 2\mu_0^2 \left[\mathfrak{n}(E_0) + \frac{\pi^2 k^2 T^2}{6} \left\{ \frac{1}{\mathfrak{n}} \frac{\partial^2 \mathfrak{n}}{\partial E^2} - \left(\frac{1}{\mathfrak{n}} \frac{\partial \mathfrak{n}}{\partial E} \right)^2 \right\}_{E=E_0} \right]. \quad (5\cdot11)$$

5·4. Diamagnetism.

The diamagnetic effect caused by the rotation of the electronic orbits due to a magnetic field is a simple problem for atoms. The susceptibility per atom is given by

$$-\frac{\epsilon^2}{6mc^2} \Sigma \overline{r^2}, \quad (5\cdot12)$$

where $-\epsilon$ and m are the charge and mass of the electron, r is the distance of an electron from the nucleus, the bar denotes the space mean value and the summation is over all the electrons. This formula can also be applied to the ionic cores of the atoms in a metal, since the inner electrons are unaffected by the presence of the neighbouring atoms and $\overline{r^2}$ is the same as in the free atom.

The theory of the diamagnetism of the free electrons is much more difficult. Formula (5·12) is obviously inapplicable since there is no unique origin from which to measure r, and (5·12) is not invariant for a change in origin. The correct theory was first given by Landau, who showed that, whereas according to the classical theory the diamagnetism of free electrons is zero, a diamagnetic effect is to be expected according to quantum theory. The difference between the classical and quantum behaviours is difficult to explain intuitively, but roughly it arises because the orbits of a free electron in a magnetic field are quantized, while there is no such restriction on the orbits in the classical theory. The motion of an electron parallel to the direction of the field H is unaffected by the field and is therefore unquantized, while the projection of the path of an electron on a plane perpendicular to H is a circle, described with angular frequency $\nu = \epsilon H / 4\pi mc$. Since a circular motion can be considered as made up of two simple harmonic motions at right angles, the possible energy levels are given by $E=(n_1+n_2+1)\,h\nu$, where n_1 and n_2 are integers, the term $h\nu$ arising because each oscillator has a zero-point energy of $\frac{1}{2}h\nu$. The magnetic moment of a stationary state is $-\partial E/\partial H=-(n_1+n_2+1)\,\epsilon h/4\pi mc$, and the average of this over all the occupied energy levels has to be calculated. Landau's result is that for perfectly free electrons the diamagnetism is exactly one-third of the spin paramagnetism, for both the Boltzmann and the Fermi-Dirac statistics.

5·41. A method for dealing with the diamagnetism of the quasi-free conduction electrons in metals was given by Peierls and a more powerful method was given by Wilson. Only the result can be given here. It is found (*T.M.* p. 118, equation (178)) that the susceptibility is mainly determined by the quantity

$$\frac{\partial^2 E}{\partial k_1{}^2}\frac{\partial^2 E}{\partial k_2{}^2}-\left(\frac{\partial^2 E}{\partial k_1 \partial k_2}\right)^2, \qquad (5\cdot 13)$$

where E is the energy as a function of the wave vector $\mathbf{k}$, and k_1, k_2 are the components of $\mathbf{k}$ at right angles to the field. The quantity (5·13) is connected with the Gaussian curvature of the energy contours in the $\mathbf{k}$ space, and it can be shown that this curvature can only be abnormally large when the energy contour comes near to a zone boundary. For perfectly free electrons the diamagnetic susceptibility is the same as that calculated by Landau, as it must be.

5·5. A survey of the magnetic properties of metals.

Before we can compare the theoretical and experimental results, we have to subtract from the observed values the diamagnetism of the atomic cores. These corrections can be estimated partly by calculating (5·12) from the known electronic distributions, and partly empirically from the susceptibilities of compounds containing the ions. Unfortunately there are two difficulties which prevent us from getting as much information as we would like from the discussion of the theoretical formulae. The first difficulty is that at present we have practically no knowledge of the curvature of the energy surfaces, on which the diamagnetic contribution depends. (The calculations mentioned in § 2·4 give the density of states but not the curvature, which requires very much more accurate calculations.) The other difficulty, which is perhaps even more serious, arises from the effect of the exchange and correlation forces between the electrons. The exchange forces tend to align the spins of the electrons and thus to increase the magnetic moment, and although this is offset to a certain extent by the effect of the correlation forces, it is nevertheless an important factor. Estimates have been made of the magnitudes of these forces to an accuracy which is sufficient for dealing with cohesion (see § 3·2), but the magnetic susceptibility is a much more delicate problem and the estimates so far given lead to impossibly large susceptibilities. The best that we can do is to neglect the effect of the exchange forces entirely and to use Landau's value for the diamagnetic susceptibility. We can also introduce an effective mass m^* different from m; in this case the diamagnetic contribution is no longer one-third the paramagnetic, since the spin magnetic moment μ_0 is independent of m^*, whereas the orbital moment is not. Hence we must replace m by m^* only in the factor $\mathfrak{n}(E_0)$ which occurs in (5·7), while we must replace every factor m by m^* in the diamagnetic term. Since

$$\mathfrak{n}(E_0) = 2\pi(2m^*)^{\frac{3}{2}} h^{-3} E_0^{\frac{1}{2}} = 2\pi m^* h^{-2} (3n/\pi)^{\frac{1}{3}}$$

for free electrons, we have

$$\chi = 4\pi \left(\frac{3n}{\pi}\right)^{\frac{1}{3}} \frac{\mu_0^2}{h^2} \left(m^* - \frac{m^2}{3m^*}\right). \qquad (5·14)$$

Alternatively, if $\mathfrak{n}(E_0)/n_a$ is the density of states per electron volt per atom (for one direction of spin), the susceptibility χ_A per gram atom is given by

$$\chi_A = 64 \times 10^{-6} \left(1 - \frac{m^2}{3m^{*2}}\right) \frac{\mathfrak{n}(E_0)}{n_a}. \qquad (5·15)$$

We see from (5·14) that a large effective mass favours the paramagnetism while a small effective mass favours the diamagnetism. This is what we should expect, since when m^* is large the energy bands are narrow and all the electrons and not only the fastest should contribute to the spin paramagnetism. On the other hand, when m^* is small the Larmor frequency of rotation $ehH/(4\pi m^* c)$ is large and hence the contribution to the diamagnetic susceptibility should be large.

The susceptibilities of the monovalent elements are fairly well represented by (5·14) with $m^* = m$, as is shown by the figures given in Table III. The calculated values of χ are uniformly too low, even allowing for the great uncertainty in the experimental values, and this is probably due to the neglect of the exchange forces, though in some cases it may be partly due to the effective mass of the electrons being greater than m. We have not given the values of $\mathfrak{n}(E_0)$ calculated from the observed values of χ, since, for the reasons given above, these cannot be compared with the values calculated from the specific heat.

TABLE III. Volume susceptibilities in units of 10^{-7}

	$n \times 10^{-21}$	Total χ observed	χ due to ions	χ due to valency electrons	χ calculated
Li	48·3	(2·8)	− 0·56	(3·36)	5·3
Na	26·2	5·0 6·5	− 2·3	7·3 8·8	4·2
K	13·8	3·6 4·9	− 3·0	6·6 7·9	3·5
Rb	11·3	1·1 3·3	− 4·2	5·3 7·5	3·3
Cs	9·0	−2·0 4·4	− 5·2	5·0 9·6	3·1
Cu	85	− 7·6	−20	13·4	6·5
Ag	59	−21	−(25 to 30)	(4 to 9)	5·9
Au	59	−29	−43	14	5·9

So far as is known, the temperature variation of χ for the monovalent metals is very small, and is thus in agreement with the theory. Measurements at high temperatures cannot be carried out for the alkalis on account of their low melting points, and measurements for the noble metals are lacking, so that it is impossible to verify even the term involving T^2 in χ.

5·51. The alkaline earths are moderately paramagnetic, the values of $\mathfrak{n}(E_0)/n_a$ calculated from (5·15) with $m^* = m$ being about 0·4 (electron volt)$^{-1}$. Barium is remarkable in that χ increases with temperature, whereas the more normal behaviour is a decrease. However, (5·11) shows that χ may increase initially with T if $\partial^2\mathfrak{n}/\partial E^2$ is positive and sufficiently large. It must of course decrease for very large T since, when the electron gas behaves classically, we have $\chi \propto T^{-1}$. If $\partial^2\mathfrak{n}/\partial E^2$ is positive, E must be such that $\mathfrak{n}(E)$ is near a minimum, and this can only happen when two bands overlap. It is therefore not surprising to find a positive temperature coefficient for χ for a divalent element. However, the order of magnitude of the effect is so large that it is difficult to believe that the above explanation is the correct one. (According to Lane(14), χ_A increases from 20×10^{-6} at 20° C. to 57×10^{-6} at 400° C.)

The susceptibility of the free electrons in the divalent elements zinc, cadmium and mercury is paramagnetic, but it is less than we should expect from the values of $\mathfrak{n}(E_0)$ found from measurements of the specific heats (Table II). This may mean that the diamagnetic contributions of the free electrons has been underestimated, but since zinc and cadmium become markedly more diamagnetic as the temperature is lowered they do not in any case fit into the theory.

Bismuth is the most diamagnetic ($\chi_A = -290 \times 10^{-6}$ at 18° C.) of all the metals, and it becomes much more diamagnetic as the temperature is lowered. This immense diamagnetism is supposed to be produced by the quasi-free electrons, which, in bismuth, nearly fill a Brillouin zone, and are therefore most favourably situated to give a large diamagnetic effect. We should, however, expect the susceptibility to be nearly constant especially at low temperatures, whereas in fact $\partial\chi_A/\partial T \sim 0{\cdot}3$. Thus although the theory is satisfactory qualitatively and indicates the cause of the peculiar magnetic properties of bismuth, it is not at present very good quantitatively. A further discussion of the properties of bismuth is given in § 6·36.

5·52. Those transition elements, such as manganese, platinum and palladium, which are not ferromagnetic are strongly paramagnetic, a value of 3 for $\mathfrak{n}(E_0)/n_a$ being required according to (5·15), if we neglect the diamagnetism, in order to account for the susceptibility of platinum at room temperature. Since this value of $\mathfrak{n}(E_0)/n_a$ is very much larger than the value found from the specific heat (Table II), we deduce that the effect of the exchange forces is very important, and that the simple theory is quite inadequate to deal with the transition elements. This conclusion is borne out by the very large temperature variation of χ. For these metals, and in particular for palladium, $\partial\chi/\partial T$ is large and negative even for liquid-air temperatures, whereas we should not expect any appreciable variation in χ for temperatures lower than a few hundreds of degrees Centigrade.

The magnetic properties of the alloys of the transition elements are of considerable theoretical importance. When copper, silver or gold is added to palladium, χ decreases and becomes zero when about 50 per cent of the noble metal has been added. Since we ascribe the largeness of the paramagnetic susceptibility to the large density of states in the *d*-band, even though the simple theory is inadequate, this means that zero susceptibility is attained when the *d*-band is full, and we therefore deduce that the number of unoccupied levels in the *d*-band in pure palladium is about 0·5 per atom. (We assume that each monovalent atom adds one electron to the *d*-band until it is full.) In the free state palladium and platinum have ten electrons in the 4*d*- and the 5*d*-states respectively, and these ten electrons are just sufficient to fill the corresponding *d*-bands in the metals. However, on account of the overlapping of the *d*-bands by the 5*s*- and 6*s*-bands in the metallic states, there is in both metals about 0·5 electron per atom in the *s*-band and the same number of vacant levels in the *d*-band.

5·6. Ferromagnetism.

Weiss' theory, which postulates the existence of an inner field tending to align the elementary magnets, gave a very satisfactory formal account of ferromagnetism, but it was not until 1928, when Heisenberg noticed that the energies involved in ferromagnetic phenomena are of the same order as the energy differences between the ortho and para states of helium, that any explanation could be given of the origin of the inner field.

The para states of helium are singlets and the spins of the electrons are anti-parallel; for these states the spatial wave function of the electrons must be symmetrical, since, according to the exclusion principle, the total wave function, including both the spatial and spin portions, must be anti-symmetrical. Now if a helium atom changes from say the 2*s* para state to the 2*s* ortho state the spins become parallel, and, to preserve the anti-symmetry of the total wave function, the spatial part must become anti-symmetrical. Therefore in corresponding para and ortho states the electronic configuration is necessarily different and there is therefore a large energy difference, which is called the exchange energy. Although the exchange energy is of an electrostatic nature, yet we see that its effect can be described formally as introducing a strong coupling between the spins of the electrons. Weiss' theory postulates such a coupling.

In his original theory, Heisenberg considered the idealized problem of a crystal composed of atoms each of which had one valency electron in an *s*-state. It is impossible to find the energy levels of such a complicated system, and Heisenberg therefore assumed that all the excited states of the electrons could be neglected, and in particular that there were no polar states in which two electrons were on the same atom. By making this assumption Heisenberg was able to find the lowest energy levels, but the possibility of conduction was necessarily excluded from the theory. A further assumption had to be made that the exchange energy is positive so that the most stable state of the metal at the absolute zero should be that in which all the spins are parallel. In most molecules and solids the exchange energy is negative.

It would take us too far afield to discuss either the formal theory of ferromagnetism or Heisenberg's theory, and we consider instead the attempts which have been made to attack the problem from the point of view of the conduction theory. It was first pointed out by Bloch that the exchange energy for free electrons is positive and hence tends to make the spins of the electrons point in the same direction. However, any alignment of the spins increases the kinetic energy of the electrons, since if the spins are parallel each energy level can only accommodate one electron instead of two. There are, therefore, three possibilities. (1) If the exchange forces are weak, the lowest state is the unmagnetized state in which each occupied level contains two electrons with opposite spins. (2) If the exchange forces exceed a certain limit the decrease in potential energy when the spin of an electron is reversed more than compensates for the increase in kinetic energy necessitated by the promotion of the electron to a previously unoccupied level. Since the energy required to reverse a spin increases with the number of spins reversed (the electrons have to go into higher and higher levels), the exchange forces may only be large enough to ensure that some but not all of the levels are occupied by electrons with one direction of spin, the other levels being occupied by two electrons with opposite spins. In this case only the singly occupied levels contribute to the magnetic moment. (3) If the exchange forces are very large the lowest state will be that in which all the electrons have the same direction of spin. It is only in case (3) that there can be any simple relation between the magnetic moment per atom—the magneton number—and the number of valency electrons per atom.

5·61. One of the serious difficulties in the way of a complete theory is that we have little knowledge of the exchange forces. In spite of this we can make several deductions of a general nature. Since the exchange forces have to overcome the increase in the kinetic energy we should expect them to be most effective when the spacing between the energy levels is small, that is, when the density of states is large. Now we have seen that the density of states is very large when there is an incomplete *d*-band, and it is just those elements which possess such a band that are either ferromagnetic or very strongly paramagnetic.

Let us discuss nickel in detail; the density of states is shown in fig. 13, p. 29, the contributions from the 3*d*- and the 4*s*-bands being shown separately. There are ten electrons to be accommodated in the 3*d*- and 4*s*-bands, and, if the exchange forces are neglected, the electrons occupy the lowest possible levels and hence fill the bands up to the energy marked 10. Most of the electrons are in the *d*-band, but some (less than one per atom) are in the *s*-band. It is not very clear exactly what the effect of the exchange forces is, and there is a considerable number of possibilities. If we assume, as we must in order to make ferromagnetism possible, that there is a net gain in energy when more electrons have their spins pointing say to the right than to the left, it is reasonable to assume that the electrons in singly occupied levels are situated in the *d*-band, since the number of electrons which can be accommodated in a given energy interval is much greater for the *d*-band than for the *s*-band. We now divide the states into two sets one of which accommodates electrons with spins pointing to the right and the other of which accommodates electrons with spins pointing to the left. These sets, of course, extend over the same energy range. The simplest assumption that we can make is that the states are filled as follows. We assume that all the states in the *d*-band for electrons with spins pointing to the right are filled, which requires five electrons per atom, and that some of the states in the *d*-band for electrons with spins pointing to the left are filled, the number of electrons of the latter type being $5-x$ per atom, where $0<x<5$. The remaining x electrons per atom occupy part of the *s*-band, each level having two electrons with opposite spins. To calculate x we need to know the exchange forces and the density of states in the two bands. A first attempt to carry out the calculations has been made by Slater (5), but it is simpler to deduce x from the experimental results. With the above assumptions the magnetic moment per atom of nickel is $x\mu_0$, since x is the number of unpaired electrons, and hence x is uniquely determined by the magnetic moment. (With other possible and plausible assumptions the number of empty levels in the *d*-band is not uniquely determined by the magnetic moment.) Since the saturation value of the magnetic moment of nickel at 0° K. is 0·6 Bohr magneton we deduce that $x=0{\cdot}6$, i.e. that there is 0·6 unoccupied level per atom in the *d*-band and 0·6 electron per atom in the *s*-band.

5·62. Ferromagnetic alloys.

According to the argument given in the preceding section there is 0·6 unpaired electron per atom in nickel occupying the highest energy levels of the $3d$-band. The corresponding numbers, deduced in the same way, are 1·7 and 2·2 for cobalt and iron respectively, i.e. there are 8·3 and 7·8 electrons per atom in the d-band. The number of electrons per atom in the $4s$-band is $9-8\cdot3=0\cdot7$ for cobalt and $8-7\cdot8=0\cdot2$ for iron. These figures are, however, not too reliable since they are based on assumptions which probably do not hold, at least for iron. It is probable that the exchange forces are not sufficiently strong to make the lowest state in iron the one in which all the energy levels in the d-band for electrons with spins pointing to the right are filled. In this case there must be $5-y$ electrons per atom in the d-band with spins pointing to the right, $5-x$ with spins pointing to the left ($x>y$) and $x+y-2$ paired electrons in the $4s$-band ($x+y>2$). The magneton number is $x-y$ Bohr magnetons, and we cannot therefore deduce both x and y from measurements of the magnetic moment alone, and we cannot find the number of electrons in the $4s$-band. Support for the hypothesis that $y\neq0$ for iron is given by the fact that the magneton number of iron is increased by the addition of small quantities of cobalt and nickel, which increase the number of electrons present. The magneton number of the ferromagnetic materials therefore reaches a maximum when the number of electrons outside the closed $3p$-shell is somewhere between 8 and 9, and it is presumably at this concentration that $y=0$.

The above discussion makes it clear that one of the important factors affecting ferromagnetism is the number of electrons in the $3d$-band, and that there is a certain optimum number round about 8·5 electrons per atom. This enables us to understand in a general way the behaviour of the ferromagnetic alloys. When copper is alloyed with nickel the magneton number is reduced, owing to the valency electrons of the copper atoms going into the $3d$-band of the alloy and compensating some of the previously unpaired spins. The extrapolated results indicate that ferromagnetism disappears when about 60 per cent of copper has been added, which confirms the hypothesis that in pure nickel there is 0·6 vacant level per atom in the 3*d*-band. The addition of zinc, aluminium and other electropositive metals has the same effect, the valency electrons of the added metal compensating some of the spins of the nickel atoms. The general behaviour is shown in fig. 24.

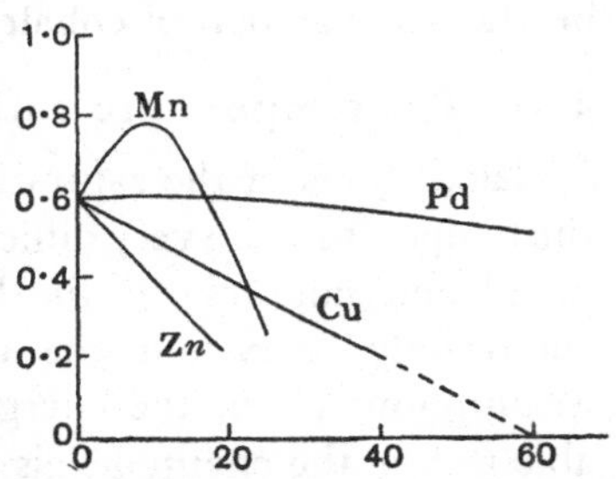

Fig. 24. The magneton numbers of nickel alloys as functions of the composition.

Not all the ferromagnetic alloys can be dealt with by merely considering the number of available electrons, just as all alloys do not obey the Hume-Rothery rules for the effect of the electron concentration. Other factors are often very important and these are very imperfectly understood at present. This is most clearly shown by the ferromagnetic alloys of manganese. Since manganese has seven electrons outside the closed 3*p*-shell, we can obtain any number of electrons between seven and eleven outside the 3*p*-shell by alloying manganese and copper. If the electron concentration alone is important, we should expect that for a certain range of concentrations these alloys should be ferromagnetic, whereas in fact they are at most very strongly paramagnetic. On the other hand the ternary alloys of manganese, copper and aluminium embrace the ferromagnetic Heusler alloys. The maximum magnetization occurs for the composition $MnAlCu_2$, and the alloys are in the ordered state. The saturation intensity is almost the same as that of nickel, and the number of electrons outside the 3*p*-shell is 10·5 per atom and is therefore also almost the same as for nickel. Thus in some respects the Heusler alloys fit into the general scheme, but it is difficult to explain why the presence of three constituents is necessary to produce ferromagnetism.

The alloys of manganese and nickel are also somewhat anomalous. The addition of manganese to nickel increases the magnetic moment for not too large concentrations, in agreement with the general theory, but the maximum magneton number is only about 0·75 Bohr magneton, whereas, if the electron concentration were the only important factor, the magneton number, when $33\frac{1}{3}$ per cent of manganese has been added, should be the same as that of cobalt.

5·63. The temperature variation of the magnetization.

Calculations of the saturation magnetization as a function of the temperature are very difficult, since they involve a knowledge of all and not only of the lowest energy levels. Heisenberg could only carry out the calculations by making arbitrary assumptions about the energy level system, but Bloch (15) was able to find the energy levels sufficiently accurately to deal with the magnetization at low temperatures. He found that, if M is the magnetization at temperature T and M_0 is the value of M at $T=0$, then

$$\frac{M}{M_0}=1-a\left(\frac{kT}{J}\right)^{\frac{3}{2}}, \tag{5·16}$$

where J is the exchange energy tending to align the spins and a is a constant depending on the crystal structure ($a=0·066$ for a body-centred cubic lattice and $a=0·033$ for a free-centred cubic lattice).

Slater (5) has recently carried out calculations of M from the point of view adopted here, and Stoner (16) has considered in detail a special model in which there are n electrons in an energy band of the normal type. Stoner does not deal with the exchange forces exactly, but introduces an arbitrary parameter to take account of their effect, which enables the calculations to be carried out. He finds that the temperature variation depends on the magnitude of the exchange forces. If the exchange forces are large enough to produce ferromagnetism, but not large enough to align all the spins even at the absolute zero, so that $M_0<n\mu_0$, the magnetization is given very closely by

$$\left(\frac{M}{M_0}\right)^2=1-\left(\frac{T}{\theta}\right)^2, \tag{5·17}$$

whereas, if the exchange forces are large enough to make $M_0=n\mu_0$, then M/M_0 approaches unity exponentially as $T\to 0$.

EFFECT OF TEMPERATURE

The temperature dependence of the magnetization and the curvature of the $1/\chi$, T-curves above the Curie point are in excellent qualitative agreement with the observations, and, although this new attack upon the theory of ferromagnetism is still in its infancy, yet it promises to be more fruitful than any other in correlating the magnetic and electrical properties of the ferromagnetic metals.

REFERENCES

(1) W. H. Keesom and J. A. Kok. *Physica* (2), **1** (1934), 770; **3** (1936), 1035 and **4** (1937), 835.
(2) W. H. Keesom and C. W. Clark. *Physica* (2), **2** (1935), 513.
(3) G. L. Pickard. *Nature*, **138** (1936), 123.
(4) J. G. Daunt and K. Mendelssohn. *Proc. Roy. Soc.* A, **160** (1937), 127.
(5) J. C. Slater. *Phys. Rev.* **49** (1936), 537 and 931.
(6) E. C. Stoner. *Phil. Mag.* **25** (1938), 899.
(7) F. M. Jaeger, E. Rosenbohm and J. A. Bottema. *Proc. Amsterdam Acad.* **35** (1932), 772.
(8) E. C. Stoner. *Phil. Mag.* **22** (1936), 81.
(9) E. Lapp. *Ann. Physique*, **12** (1929), 442.
(10) K. E. Grew. *Proc. Roy. Soc.* A, **145** (1934), 509.
(11) H. Klinkardt. *Ann. Physik* (4), **84** (1927), 67.
(12) C. Sykes and H. Wilkinson. *Proc. Phys. Soc.* **50** (1938), 834.
(13) E. C. Stoner. *Proc. Roy. Soc.* A, **154** (1936), 656; *Proc. Leeds Phil. Soc.* **3** (1938), 403.
(14) C. W. Lane. *Phys. Rev.* **44** (1933), 43.
(15) F. Bloch. *Zeit. Phys.* **61** (1930), 206.
(16) E. C. Stoner. *Proc. Roy. Soc.* A, **165** (1938), 372.

General references

J. H. van Vleck. *Electric and magnetic susceptibilities.* (Oxford, 1933.)
F. Bloch. "Molekulartheorie des Magnetismus", *Handbuch der Radiologie*, 2nd ed., vol. 6, part 2. (Leipzig, 1934.)
E. C. Stoner. *Magnetism and matter.* (Methuen, 1934.)
J. McDougall and E. C. Stoner. "The computation of Fermi-Dirac functions", *Phil. Trans.* A, **237** (1938), 67.
E. Vogt. "Magnetismus der metallischen Elemente", *Ergebnisse der exakten Naturwissenschaften*, **11**. (Berlin, 1932.)

Chapter VI

CONDUCTIVITY

6·1. The time of relaxation.

The conductivity of metals is much more difficult to deal with theoretically than any of the equilibrium properties, since not only do we need to know the exact wave functions and energy levels (a knowledge of the density of levels is not sufficient) but we meet with great mathematical difficulties. In transport problems we do not know the velocity distribution of the electrons *a priori*; the effect of external fields and temperature gradients is to tend to modify the distribution function f, while the interaction of the electrons with the metal lattice tends to restore the distribution function to its equilibrium form f_0. A steady state is obtained when these two opposing tendencies cancel, and the determination of the distribution function depends upon the solution of an integral equation, which has so far only been carried out in some simple special cases.

The calculation of transport phenomena can be greatly simplified by use of the time of relaxation τ, which is the average time between two collisions of an electron with the lattice. It is well known from the dynamical theory of gases that it is impossible to define the average time between collisions uniquely, and it is therefore preferable to define τ as follows. Consider any non-equilibrium state for which the distribution function is f. If there are no external fields, it is a plausible assumption that the rate at which equilibrium is restored is proportional to the difference $f-f_0$. Thus, if $[\partial f/\partial t]_{\text{coll}}$ is the rate of change of f, we put

$$\left[\frac{\partial f}{\partial t}\right]_{\text{coll}} = -\frac{f-f_0}{\tau}, \tag{6·1}$$

where τ is the time of relaxation. Hence the approach to equilibrium follows the equation $f-f_0 = \text{constant} \times e^{-t/\tau}$. The definition (6·1) must be used in the quantitative theory, but in the descriptive treatment given here it is simpler to use the first definition of τ. We further assume that, unless otherwise stated, the electrons are to be treated as free so that the energy is simply $\frac{1}{2}mv^2$. The calculations which follow are extremely rough and are only given so as to indicate the order of magnitude of the effects and to show what are the important factors. In general we consider a one-dimensional model whenever no important feature is lost by ignoring the velocity components at right angles to the external field.

6·2. The electrical conductivity.

When a uniform electric field $\mathscr{E}$ is present, the electrons have an acceleration $-\epsilon\mathscr{E}/m$, where $-\epsilon$ is the charge of the electron. An electron whose equilibrium velocity is v acquires an extra velocity $-\epsilon\mathscr{E}\tau/m$ in the time τ between two collisions. If we assume that this extra velocity is lost during a collision, the average velocity of an electron is $v-\frac{1}{2}\epsilon\mathscr{E}\tau/m$, and so the effect of the field is to increase the velocity of each electron by $-\frac{1}{2}\epsilon\mathscr{E}\tau/m$. This is illustrated by fig. 25, where $f(E)$ is given as a function† of v, which takes both positive and negative values; the continuous line gives $f_0(E)$ and the broken line $f(E)$ obtained by shifting $f_0(E)$ a distance $-\frac{1}{2}\epsilon\mathscr{E}\tau/m$ along the v axis. Since the currents of the electrons with velocities v and $-v$ cancel, we see that the only electrons which contribute to the resultant current are those for which $f(E)$ and $f_0(E)$ are different. The current is therefore due to the excess of electrons with velocities near v_0 moving in one direction and the deficiency of electrons with velocities near $-v_0$ moving in the other direction. ($|\,v_0\,|$ is

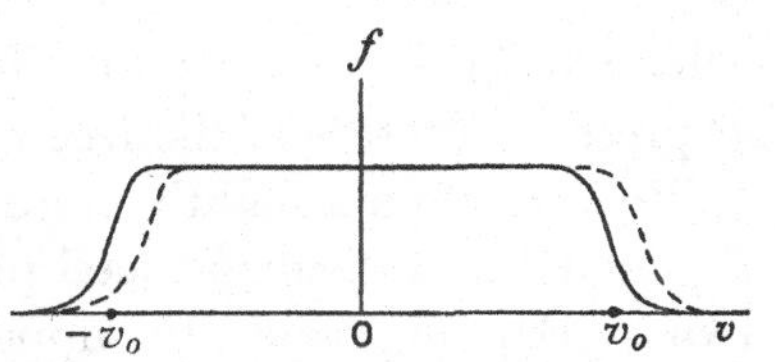

Fig. 25. The velocity distribution function. The continuous curve gives the equilibrium distribution function and the broken curve the distribution function in the presence of an electric field.

† Note that $f(E)$ is the average number of electrons occupying *one* energy level and not the number of electrons occupying an energy range.

the maximum velocity of the electrons at the absolute zero and is given by $E_0 = \frac{1}{2}mv_0^2$.)

The electrons with velocities lying between v_0 and $v_0 - \frac{1}{2}\epsilon\mathscr{E}\tau/m$ occupy an energy interval ΔE given by

$$\Delta E = -\tfrac{1}{2}mv_0^2 + \tfrac{1}{2}m(v_0 - \tfrac{1}{2}\epsilon\mathscr{E}\tau/m)^2 = -\tfrac{1}{2}\epsilon\mathscr{E}\tau v_0 + O(\mathscr{E}^2).$$

The total number of electrons moving to the right is $\frac{1}{2}n$ per unit volume, and so the number lying in the energy interval ΔE is of the order $\frac{1}{2}n\Delta E/E_0 = -\frac{1}{4}n\epsilon\mathscr{E}\tau v_0/E_0$. The velocity of these electrons is v_0 approximately, so that the rate of transport of charge per unit area is $\frac{1}{4}n\epsilon^2\mathscr{E}\tau v_0^2/E_0$. Since there is a deficiency of electrons moving to the left equal to the excess moving to the right, the resultant current density is

$$\tfrac{1}{2}n\epsilon^2\mathscr{E}\tau v_0^2/E_0 = n\epsilon^2\mathscr{E}\tau/m.$$

We thus obtain for the conductivity σ the expression

$$\sigma = n\epsilon^2\tau(E_0)/m, \tag{6·2}$$

where $\tau(E_0)$ is the value of τ for $E = E_0$. We can, if we wish, express σ in terms of the free path l by defining $l = \tau v$.

We can obtain a slightly more general formula by considering the model of a divalent metal proposed in § 2·51, consisting of two overlapping bands, the upper of which contains n_2 electrons per unit volume and the lower of which contains n_1 holes. We then have

$$\sigma = \epsilon^2\left(\frac{n_1\tau_1}{m_1} + \frac{n_2\tau_2}{m_2}\right). \tag{6·3}$$

6·21. The ideal and residual resistances.

Since an electron can move freely through a perfect lattice, the resistance must be caused by the departure of the lattice from perfect regularity. This can arise in two ways, (1) by imperfections caused by cracks, strains and the presence of foreign atoms, and (2) by the thermal motion of the lattice. Since these two mechanisms of scattering are independent of one another, the total probability of an electron being scattered is the sum of the separate probabilities of its being scattered by the mechanisms (1) and (2). Hence, if τ is the total time of relaxation and if τ_0 and τ_i are the partial times of relaxation due to the mechanisms (1) and (2) respectively, then, by (6·1), we have

$$\frac{1}{\tau}=\frac{1}{\tau_0}+\frac{1}{\tau_i},$$

and, by (6·2), we can write

$$\frac{1}{\sigma}=\frac{1}{\sigma_0}+\frac{1}{\sigma_i}, \qquad (6·4)$$

where $1/\sigma_0$ is the "impurity resistivity" and $1/\sigma_i$ is the ideal electrical resistivity due to the heat motion of the lattice.

The free path for collisions of the electrons with the imperfections must be of the order of the average distance between the imperfections and is therefore independent of the velocity and the temperature. The corresponding resistance must also be independent of the temperature and it will vary from specimen to specimen. The interaction of the electrons with the lattice, on the other hand, takes place mainly through the scattering of the electrons by the fluctuations in density caused by the thermal vibrations. Now only the longitudinal waves in a solid are associated with volume changes, the transverse waves being distortional but not compressional waves, and thus the longitudinal sound waves are the effective scatterers. The free path is therefore of the order of the mean wave length of the sound waves, and it must increase as the temperature decreases, since only the long waves are excited at low temperatures. The corresponding resistance is called the ideal resistance since it is characteristic of the pure metal; it tends to zero as $T \to 0$, and thus the total resistance at low temperatures tends to a limiting value which is entirely due to the presence of impurities and imperfections. For this reason the "impurity resistance" is called the residual resistance. To find the ideal resistance experimentally the measurements must be carried to low temperatures so as to obtain the residual resistance; by subtracting this from the total resistance we obtain the ideal resistance.

6·22. The magnitude of the ideal resistance.

The calculation of τ is, even in the simplest cases, of extreme complexity. The result contains a number of constants, the calculation of some of which is still a matter of controversy (see for example [1]). We therefore leave aside any detailed discussion of the magnitude of the conductivity, merely remarking that for silver at 18° C. the free path is $5{\cdot}2 \times 10^{-6}$ cm., and confine ourselves to some general remarks.

It might appear at first sight that multivalent metals ought to be better conductors than monovalent metals since they have more free electrons. This is not so, owing to the effect of the zone structure. (Since the conductivity is a function of Θ/T, where Θ is the Debye temperature, it is essential to compare the conductivities of metals at corresponding temperatures, measured in the scale of the respective Θ's.) In a multivalent metal the free electrons occupy more than one zone, and according to (6·3) the conductivity is determined by n_1 and n_2, the numbers of holes and electrons per unit volume in the lower and upper zones. Now n_1 and n_2 may be quite small compared with the number of atoms per unit volume. In fact the effective number of electrons in multivalent metals is less than in monovalent metals, so far as the conductivity is concerned, and monovalent metals are the best conductors.

The smallness of the effective number of electrons is undoubtedly the reason why the divalent elements are poorer conductors than the monovalent elements. The extreme example is bismuth, in which the effective number of electrons is of the order of 10^{-3} per atom (§§ 6·25 and 6·36). Mott [2] has, however, suggested that in the transition and ferromagnetic metals the low conductivity is due rather to an abnormal smallness in the free path than to the small effective number of electrons. These elements possess uncompleted *d*-bands, platinum, for example, having about 0·5 electron per atom in the 6*s*-band and the same number of holes in the 5*d*-band (§ 5·53). Hence the *s*-electrons can make transitions to the vacant *d*-states in addition to the normal transitions to the vacant *s*-states, which are alone possible in a monovalent metal. Thus the scattering probability is abnormally high in the transition metals and the free path is abnormally low.

6·23. The temperature variation of the ideal resistance.

In order to determine the temperature variation of the ideal resistances it is only necessary to know how τ depends on the temperature. We see from (6·1) that $1/\tau$ is proportional to the transition (i.e. scattering) probability of an electron from one state of motion to another, since this determines $[\partial f/\partial t]_{\text{coll}}$. Now it is a well-known result of the perturbation theory of quantum mechanics that a transition probability is proportional to the square of the matrix element of the perturbation causing the transition. In the problem which we are considering, the perturbation is proportional to the amplitude of the sound waves. Hence $1/\tau$ is proportional to the square of the amplitude, i.e. to the energy of the sound waves, which is proportional to T at temperatures much greater than the Debye temperature. We therefore have finally the result that the resistance is proportional to T at high temperatures. This result ceases to hold at low temperatures since we have neglected a number of important factors, chief among which is that the electrons are reflected by the sound waves with a change in momentum, the change being of the nature of a Doppler effect. It is found that (*T.M.* p. 213, equation (343))

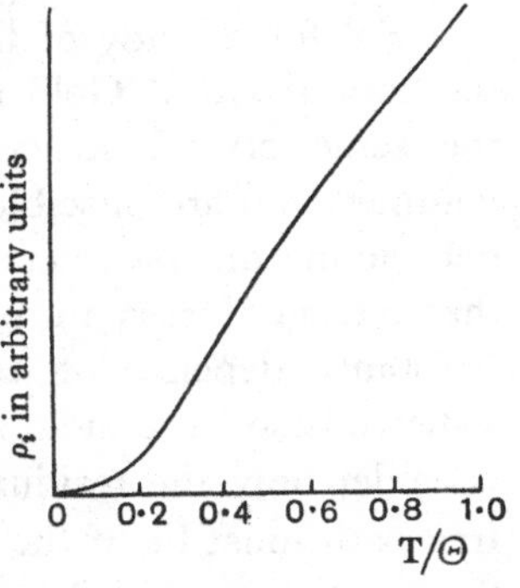

Fig. 26. The ideal resistance as a function of the temperature.

$$\rho_i \propto \left(\frac{T}{\Theta}\right)^5 \int_0^{\Theta/T} \frac{z^5 dz}{(e^z - 1)(1 - e^{-z})}, \tag{6·5}$$

where ρ_i is the ideal resistivity and Θ is the Debye temperature. The integral can readily be evaluated when Θ/T is either very large or very small, and we have

$$\rho_i = \text{constant} \times 124(T/\Theta)^5, \quad (T \ll \Theta), \tag{6·6}$$

and

$$\rho_i = \text{constant} \times \tfrac{1}{4}T/\Theta, \quad (T \gg \Theta). \tag{6·7}$$

Formula (6·5) agrees very well with the experimentally determined temperature variation of the resistance. In fig. 26, ρ_i is shown graphically.

6·24. The resistance of alloys.

The addition of one metal to another can alter the resistance in three ways, (1) by changing the number of valency electrons, (2) by changing the average field acting on the valency electrons, (3) by providing scattering centres when the alloy is in the disordered state. The effect of (1) and (2) can only be discussed in relation to the energy levels of the alloy and we therefore confine our attention to (3). The gold-silver alloys provide ideal material for a study of the influence of the composition upon the conductivity. Gold and silver are both monovalent, have the same crystal structure and practically identical atomic volumes, and are miscible in all proportions. Since the average field acting on the valency electrons is the (weighted) mean of the average forces in pure silver and pure gold, the ideal resistance depends on the composition, but the residual resistance is so large in an alloy that we can ignore this effect and consider only the residual resistance. For dilute solutions the free path must be of the order of the average distance apart of the solute atoms, and thus the additional resistance caused by alloying must be proportional to concentration of solute atoms. For large concentrations of both components the resistance is proportional to c_1c_2, where c_1, c_2 are the concentrations of the components. This can be seen as follows. Since the resistance is proportional to the collision probability, which in turn is proportional to the square of the perturbing energy causing the transitions, the extra resistance must be a quadratic function of the concentrations. Further, the extra resistance vanishes when either c_1 or c_2 vanishes and hence it must be proportional to c_1c_2. The resistances of the gold-silver alloys are shown in fig. 27, the curve for 0° K. being, of course, extrapolated. The results are in good agreement with the theory.

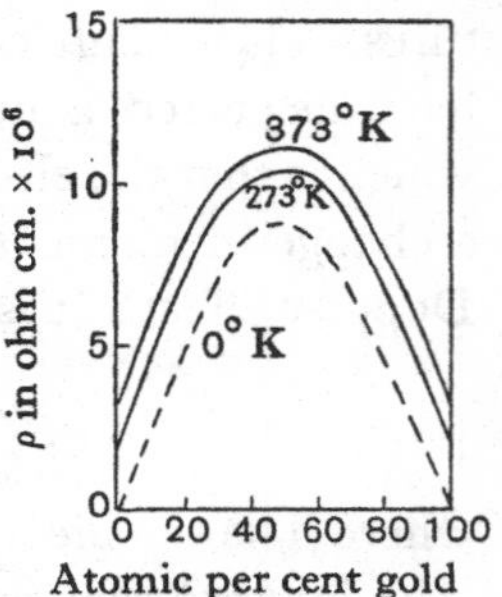

Fig. 27. The specific resistance of the silver-gold alloys.

6·25. The resistance of the bismuth alloys.

We saw in § 3·5 that the valency electrons in bismuth are just able to fill a Brillouin zone completely, but that, owing to the comparative smallness of the energy discontinuities, some electrons overlap into the next zone and leave a number of holes in the almost filled band. The number of free electrons is, therefore, so small that it can be altered appreciably by the addition of suitable impurities. It has indeed been known for a long time that the resistance of bismuth is very sensitive to the presence of small amounts of impurities, but the early results were discordant and no very reliable information could be obtained from them. A recent investigation by Thompson (3) has, however, clarified the situation considerably.

The resistance of perfectly pure bismuth increases with the temperature in the normal manner. When small amounts of lead are added, humps appear in the resistance curve, and when there is a sufficient amount of lead present the resistance curve possesses a maximum, the position of which moves to higher temperatures as the lead content increases. The general behaviour is shown in fig. 28. The interpretation of these results is as follows. Since lead has fewer valency electrons than bismuth, the replacement of an atom of bismuth by an atom of lead in the crystal reduces the number of electrons in the upper Brillouin zone. If sufficient lead is added there will be no electrons in the upper Brillouin zone, but there will be some vacant levels in the lower zone. The resistance of such an alloy behaves normally at low temperatures—the resistance increases with the temperature since the free path decreases—but, when the temperature is such that the

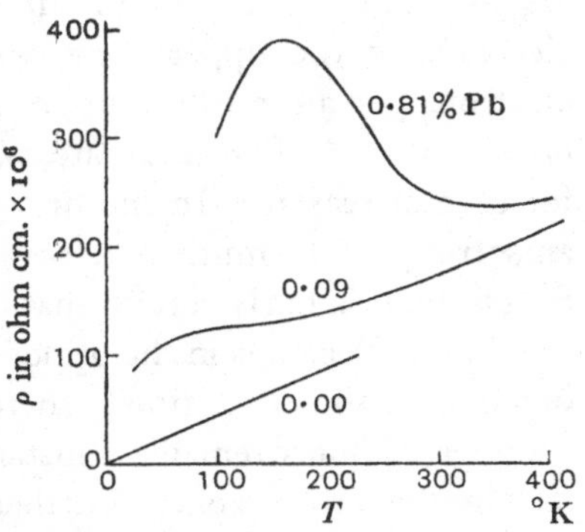

Fig. 28. The specific resistance of bismuth-lead alloys parallel to the principal axis.

thermal energy is of the same order as the energy required to raise an electron to the upper zone, the resistance must decrease, since the decrease in the free path will be more than offset by the increase in the number of electrons excited, and of the number of holes produced, as the temperature is raised. When the number of holes is increased by the addition of more lead, the energy required to excite an electron is increased, and thus the excitation of the electrons only becomes apparent at a higher temperature, which is what is observed.

The resistance parallel to the principal axis is influenced much more by impurities than is the resistance perpendicular to the axis. For example, 0·1 per cent of lead is sufficient to produce a maximum in $\rho_{\parallel}$, while 1 per cent is required to produce a maximum in $\rho_{\perp}$. Since the average resistance is $\frac{1}{3}\rho_{\parallel}+\frac{3}{2}\rho_{\perp}$, about 0·7 per cent of lead would be required to produce a maximum in the resistance of polycrystalline bismuth, though the exact figure depends on the magnitudes of $\rho_{\parallel}$ and $\rho_{\perp}$. If we assume that the maximum first appears when there are no electrons in the upper zone, we deduce that the number of electrons per atom in the upper zone of pure bismuth is of the order 7×10^{-3}. This estimate is, however, probably too large for several reasons. In the first place we have disregarded the anisotropy of bismuth and considered the average resistance. Secondly, it is fairly certain that the addition of lead reduces the number of electrons in the upper zone and increases the number of holes at the same time, whereas we have assumed that the number of holes remains constant until there are no electrons left in the upper zone. Further, tin is about three times as efficacious as lead in affecting the resistance. In view of all these factors, a figure of about 10^{-3} electron per atom is a reasonable estimate for polycrystalline bismuth.

Impurities such as selenium, which have more than five valency electrons per atom, have quite a different effect on the resistance. In these alloys the resistance behaves normally as regards temperature variation, while the first trace of impurity decreases the resistance. This is what we should expect, since the addition of electrons cannot produce the state of affairs which must exist if the resistance as a function of the temperature is to have a maximum, namely that there should be no electrons in the upper zone. When a small amount of selenium is added, the extra electrons presumably go into the upper zone and hence the conductivity is increased, while when more is added the holes begin to be filled up and the conductivity decreases.

6·3. The galvanomagnetic effects.

The most important phenomena connected with the influence of a magnetic field on the conductivity are the Hall effect and the magneto-resistance effect. Consider a rectangular metal plate in the plane $z=0$, with its edges parallel to the x- and y-axes, forming part of an isothermal electric circuit, and let there be an external electric field $\mathscr{E}_x$ along the x-axis. Let the current per unit cross-section of the plate be $J_x{}^0=\sigma_0\mathscr{E}_x$. Now suppose that there is a uniform magnetic field H along the z-axis, and suppose that the experimental arrangement is such that no transverse current, i.e. in the y direction, can flow. Then it is found that there is a transverse electric field $\mathscr{E}_y$; this is the Hall effect. Further, the current is no longer $J_x{}^0$ but some smaller quantity J_x. The conductivity σ, defined by $\sigma=J_x/\mathscr{E}_x$, is therefore a function of the magnetic field; this is called the transverse magneto-resistance effect. The conductivity is also altered when the magnetic field is along the x-axis, i.e. when the magnetic field is parallel to the current; this is known as the longitudinal magneto-resistance effect. In both cases the electrical resistance is increased by the presence of the magnetic field.

6·31. The Hall effect.

The effect of the magnetic field is that the paths of the electrons are curved between collisions instead of being straight. There is, therefore, a tendency for a transverse current to flow. If the circuit in the y direction is open, the edges of the plate must charge up until the transverse field $\mathscr{E}_y$ set up is sufficient to counterbalance on the average the effect of the magnetic field. The Hall coefficient R is defined by

$$R=\frac{\mathscr{E}_y}{J_xH}, \tag{6·8}$$

and the sign of R is taken as positive when the transverse potential is as shown in fig. 29.

Consider an electron moving with velocity v parallel to the x-axis. The force exerted on it by the magnetic field H is $\epsilon vH/c$ parallel to the y-axis (the electronic charge is $-\epsilon$), and in order to counterbalance this a transverse electrical force vH/c is required. Now if all the electrons have the same velocity we have $J_x = -n\epsilon v$, where n is the number of electrons per unit volume, and so

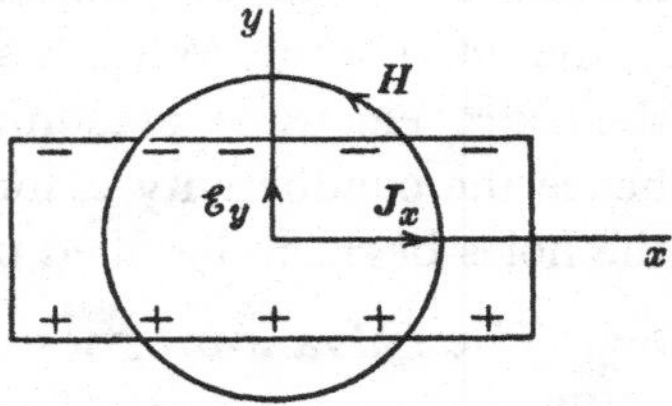

Fig. 29. The sign convention for the Hall effect. H is upwards.

$$R = \frac{vH/c}{-n\epsilon vH} = -\frac{1}{n\epsilon c}. \tag{6·9}$$

The assumption that all the electrons have the same velocity is a good approximation in metals, since only the fastest electrons take part in the conductivity, but it is not valid in say semiconductors where the electrons have a Maxwellian distribution. In this case it is impossible for the transverse field to counterbalance the effect of the magnetic field for each electron individually, and so instead $\mathscr{E}_y$ is such that the total transverse current is zero. Hence $\mathscr{E}_y = \bar{v}H/c$, where $\bar{v}$ is a certain average velocity. Also $J_x = -n\epsilon\bar{v}$, but this $\bar{v}$ is not necessarily the same as the one occurring in $\mathscr{E}_y$. The ratio of the two $\bar{v}$'s is a pure number of the order unity, and it can be shown (*T.M.* p. 171, equation (247)) that for semi-conductors

$$R = -\frac{3\pi}{8n\epsilon c}, \tag{6·10}$$

while for metals (6·9) is the appropriate formula.

Since the Hall coefficient depends on the first power of the electronic charge, there is the possibility, as pointed out in § 1·22, that it can be negative in some metals and positive in others. In order that the anomalous, positive, sign should occur it is necessary that the electrons with the greatest energy should occupy levels which lie in the region where the velocity decreases as the wave number increases. In general this means that the electrons must nearly fill a band. The Hall coefficient is then given by (6·9) or (6·10) but with the sign changed and with n now meaning the number of unoccupied levels or holes.

The general formula for a multivalent metal in which the electrons partly occupy two bands, the lower of which contains n_1 holes while the upper contains n_2 electrons per unit volume, is (*T.M.* p. 165, equation (238) for the particular case $n_1=n_2$, $\tau_1=\tau_2$)

$$R=\frac{1}{\epsilon c}\frac{n_1\tau_1^2/m_1^2-n_2\tau_2^2/m_2^2}{(n_1\tau_1/m_1+n_2\tau_2/m_2)^2}. \qquad (6\cdot11)$$

By taking $n_1=0$ or $n_2=0$ we obtain the particular cases discussed above.

6·32. Experimental results.

When the current is carried by electrons in one band only, the Hall coefficient gives at once the effective number of electrons or holes. Further, since by (6·2) and (6·9) we have

$$R\sigma=\pm\epsilon\tau(E_0)/mc,$$

the value of τ/m can be obtained. Unfortunately it is only for the monovalent elements that we can expect the hypothesis to be true; for other metals σ and R are given by formulae which reduce, when the metal is treated as isotropic, to (6·3) and (6·11). There are too many arbitrary parameters in the latter formulae for them to be determined merely by measurements of R and σ, but some of these parameters can be determined in other ways. For example, when the number of valency electrons is just sufficient to be able to fill the lower band completely (an example is a divalent cubic metal), the number of holes in the lower band must be equal to the number of electrons in the upper band. Hence $n_1=n_2$, and the number of parameters is reduced by one.

There have been many attempts to find all the parameters by combining various measurements. The most recent of these is by Ariyama (4) who made the simplifying assumption that $n_1=n_2$, $\tau_1=\tau_2$, and determined the remaining parameters from the electrical conductivity, the Hall effect and the change of electrical resistance by a magnetic field. The results are not unreasonable, but it is impossible, from the formulae at present known, to obtain correctly all the properties of any metal including say the specific heat, the magnetic susceptibility, the electrical conductivity, the Hall coefficient, the change of resistance in a magnetic field, the Thomson coefficient and the refractive index. It is in fact only in exceptional cases such as the alkalis that we can reproduce correctly the magnitudes of more than two or three of these effects, but it is also true to say that by choosing the parameters properly we can reproduce the magnitudes of any particular combination of say three quantities. We illustrate this in § 6·34 by a discussion of bismuth. Of course it is known that the theories of the optical constants and the change of resistance by a magnetic field are much more unsatisfactory than the theory of say the specific heat, because the theories of the former are not tolerably good even for the monovalent elements, yet it is difficult to say which effects ought to be given most weight. In general, we may say that if a parameter can be determined unambiguously by the measurement of one quantity then its value is likely to be correct. If it depends upon the combination of two or more measurements, the accuracy of the determination is very considerably reduced.

In view of the above discussion, we regard (6·11) as being correct qualitatively and also as regards order of magnitude, but incapable of giving exact information. We prefer, therefore, to give in Table IV the values of n, the number of electrons or holes, deduced by assuming that (6·9) is correct. The values of n_a, the number of atoms per unit volume, and of $R\sigma$, are also given. For the alkalis the values of n_a and n agree very well, as we should expect, while for the noble metals the agreement is not so good. The values of n for the other metals must be used with considerable caution. We might, for example, be inclined to imagine that in zinc there are about 1·2 positive holes per atom, a figure quite impossible to reconcile with the known zone structure. We saw in § 3·42 that zinc possesses a Brillouin zone capable of holding about 1·8 electrons per atom, and thus we should expect the number of holes in the first zone and the number of electrons in the second zone to be a few tenths of an electron per atom. The explanation of the discrepancy is as follows. According to (6·9) a small (positive or negative) Hall coefficient necessitates a large value of n, whereas in fact a small Hall coefficient can quite well arise for small values of

TABLE IV. Hall coefficients and conductivities at 0° C. in Gaussian units

	$n_a \times 10^{-22}$	$R \times 10^{25}$	$\sigma \times 10^{-16}$	$R\sigma \times 10^{7}$	$n \times 10^{-22}$ calculated from (8·9)
Li	4·8	−19	10·5	−2·0	3·7
Na	2·6	−28	20·8	−5·8	2·5
K	1·4	−47	13·3	−6·2	1·5
Rb	1·1	—	7·7	—	—
Cs	0·9	−87	5·0	−4·3	0·8
Cu	8·5	− 5·5	57	−3·2	12
Ag	5·9	− 9	60	−5·4	8
Au	5·9	− 8	44	−3·5	9
Zn	6·7	+ 8	16·3	+1·3	9
Ce	2·9	+21	1·3	+0·27	3
Ta	5·7	+11	6·5	+0·7	6
Mo	6·4	+14	21	+2·9	5
W	6·4	+13	18	+2·3	5
Ir	7·1	+ 4	18	+0·7	17
Pd	6·9	− 9	9	−0·8	8
Pt	6·7	− 2	9·2	−0·2	(35)
As	4·6	+500	2·6	+13	0·14
Sb	3·4	2000	2·5	+50	$3{\cdot}5 \times 10^{-2}$
Bi	2·8	−60000	0·9	−540	$1{\cdot}2 \times 10^{-3}$

Note. The values of R from various sources differ widely. The values given are mean values.

n_1 and n_2 by the effects of the electrons and the positive holes nearly cancelling one another (see equation (6·11)). Thus when (6·9) yields an unreasonably large value of n, it merely means that the current is carried both by electrons and by holes and that each type of carrier has about the same importance.

The values of $R\sigma$ do not have any absolute significance except for the monovalent metals, but they serve to show that the time of relaxation does not vary as much from metal to metal as we might expect.

6·33. The transverse magneto-resistance effect.

In addition to the transverse electric field, the magnetic field produces an increase in the electrical resistance which is proportional to H^2 for small fields. The model used to derive the Hall coefficient is quite inadequate to deal with this more subtle problem, since, if all the electrons have the same velocity, the change of resistance vanishes. The reason for this is as follows. The transverse electric field set up has the right magnitude to make the transverse current vanish. Therefore it is such as to counteract the average force exerted on the electrons by the magnetic field. But, if the electrons all have the same velocity, the force on each electron is the same, and hence the effect of the magnetic field and of the transverse electric field cancel exactly for each individual electron. Hence in order for there to be any effect other than the production of the Hall E.M.F. it is necessary that the electrons should not all have the same velocity. Further, it is clear that the average deviation of the electrons from the direction of the current is zero, but there is a mean square deviation and hence the change in resistance must be proportional to H^2 for small fields.

If we assume that the electron gas in a metal is completely degenerate then the only electrons taking part in conduction are the most energetic ones, and thus there is no magneto-resistance effect if the electrons are treated as being entirely free. However, if we take into account the fact that the energy range in which the conduction electrons lie is of the order kT, we obtain a change of resistance which comes out to be proportional to T^2. This is in complete disagreement with the experimental facts, since the observed change in resistance increases rapidly as the temperature is lowered. The correct theory is one which gives a magneto-resistance effect even when the electron gas is treated as completely degenerate, and to obtain this we have to give up the assumption that the electrons can be treated as free and use a model in which the energy of an electron is not a function of the magnitude of the velocity only.

The simplest model for which a magneto-resistance effect

exists is the one in which there are two overlapping bands, which we have used to describe divalent metals. It is found that

$$\frac{\Delta\rho}{\rho}=BH^2=\frac{\epsilon^2H^2}{c^2}\frac{n_1\tau_1n_2\tau_2}{m_1m_2}\frac{(\tau_1/m_1+\tau_2/m_2)^2}{(n_1\tau_1/m_1+n_2\tau_2/m_2)^2}. \quad (6\cdot12)$$

(This formula can be found from *T.M.* p. 166, equation (239), though it is not given explicitly there.)

6·34. In a semi-conductor the electrons are few in number and have a Maxwellian distribution. In this case it is impossible to assume that the electrons all have the same velocity, and hence the difficulty mentioned in the preceding section does not arise;

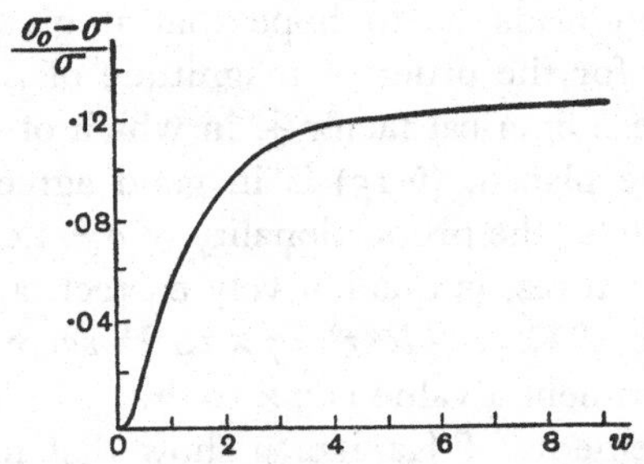

Fig. 30. The change in electrical resistance of a semi-conductor in a magnetic field. $w=\epsilon Hl/\{c(2\pi mkT)^{\frac{1}{2}}\}$.

we obtain a change in the resistance even if we treat the electrons as free. It is found (*T.M.* p. 171) that for small H

$$\frac{\Delta\rho}{\rho}=0\cdot11\frac{H^2\epsilon^2l^2}{mc^2kT}, \quad (6\cdot13)$$

where l is the mean free path. For large values of H, deviations from the quadratic law occur and finally the resistance becomes constant. If σ is the conductivity in the presence of a field and σ_0 the conductivity in zero field, then (*T.M.* p. 172)

$$\lim_{H\to\infty}\frac{\rho-\rho_0}{\rho_0}=\lim_{H\to\infty}\frac{\sigma_0-\sigma}{\sigma}=0\cdot13. \quad (6\cdot14)$$

In fig. 30 we show $(\sigma_0-\sigma)/\sigma$ as a function of w, where

$$w=\epsilon Hl/\{c(2\pi mkT)^{\frac{1}{2}}\}.$$

6·35. Any detailed comparison between theory and experiment is out of the question since the model used in deriving (6·12) is too specialized, and because, in particular, we have no simple model which is applicable to monovalent metals. We can, however, make some general observations. If we put $n_1 = n_2 = n$ and $\tau_1/m_1 = \tau_2/m_2 = \tau/m$, then (6·12) becomes, with the help of (6·3),

$$\frac{\Delta\rho}{\rho} = \frac{H^2\sigma_0^2}{4n^2\epsilon^2c^2}. \tag{6·15}$$

The same formula holds for semi-conductors except that the factor $\frac{1}{4}$ has to be replaced by 0·38. Now although (6·15) has only been derived by making very special assumptions, the form of the equation leads us to hope that it gives a reasonable approximation for the order of magnitude of $\Delta\rho/\rho$ in general. Apart from the numerical factor $\frac{1}{4}$, in which of course no trust can possibly be placed, (6·15) is in good agreement with the experimental facts; the proportionality to σ_0^2, i.e. to T^{-2} for not too low temperatures, is qualitatively correct, and the value of B for silver at 0° C. is $\frac{1}{4}R^2\sigma^2 = 7 \times 10^{-14}$ according to (6·15), while the experimental value is 3×10^{-13}.

The measurements of Kapitza (5) show that for fields of the order of a few kilogauss the change in resistance is no longer proportional to H^2, and it seems probable that this is the first sign of a flattening of the $\Delta\rho$, H curve, but this has not been observed for metals. Unfortunately it is only for semi-conductors that a reasonable theory exists for strong fields, and for these substances there are no measurements. One striking feature is that according to (6·14) the limiting value of $\Delta\rho/\rho_0$ should be 0·13 for all semi-conductors at all temperatures, and it would be very interesting to see whether this prediction is borne out or not. The only cases of apparent saturation found by Kapitza were for impure specimens of the semi-metals germanium and tellurium. For these the limiting values of $\Delta\rho/\rho_0$ were about 0·3, whereas much larger values were obtained for pure specimens of germanium and no approach to saturation was found. However, even impure germanium must be treated as a semi-metal rather than a semi-conductor, and so the above results merely hold out the hope that the theory is correct for semi-conductors, and neither definitely confirm nor disprove the theory.

There is also a change of resistance when the magnetic field is parallel to the current, but this longitudinal effect is smaller than the transverse effect. No quantitative theory has been given of this effect up to the present.

6·36. The properties of bismuth.

In order to show the kind of result which can be obtained by applying the crude theoretical formulae, we find quite formally the values of the parameters for bismuth. Since in this case $n_1 = n_2 = n$, both τ_1/m_1 and τ_2/m_2 can be found from the expressions

$$R\sigma = \frac{\epsilon}{c}\left(\frac{\tau_1}{m_1} - \frac{\tau_2}{m_2}\right), \tag{6·16}$$

and

$$B = \frac{\epsilon^2}{c^2}\frac{\tau_1 \tau_2}{m_1 m_2}. \tag{6·17}$$

The experimental values at 0° C. for polycrystalline bismuth are $R\sigma = -5{\cdot}4 \times 10^{-5}$ e.s.u. and $B = 1{\cdot}3 \times 10^{-9}$. Then (6·16) and (6·17) give $\tau_1/m_1 = 10^{15}$ and $\tau_2/m_2 = 4{\cdot}5 \times 10^{15}$. Further, since $\sigma = 9 \times 10^{15}$ e.s.u., (6·3) gives $n = 7 \times 10^{18}$. Thus the present calculations give the number of conduction electrons as $2{\cdot}5 \times 10^{-4}$ per atom, whereas the behaviour of the alloys of bismuth indicates that the number is considerably larger than this.

We cannot determine the effective masses without some independent evidence about the time of relaxation. If we assume that τ_1 and τ_2 are both about the same as τ for rubidium (we choose rubidium since the Debye Θ is of the same order of magnitude for rubidium and bismuth), we find $m_1/m = 3{\cdot}5 \times 10^{-2}$ and $m_2/m = 1{\cdot}3 \times 10^{-2}$.

These values of n, m_1 and m_2 give the right order of magnitude for the diamagnetic susceptibility if we use Landau's formula (5·13) and add the contributions from both bands; the calculated value is $\chi = -1{\cdot}4 \times 10^{-6}$ per unit volume as against the observed value of -13×10^{-6} at room temperature.

The corresponding data and results for arsenic and antimony are as follows. Arsenic, $R\sigma = +1\cdot3 \times 10^{-6}$ e.s.u.; $B = 1\cdot6 \times 10^{-11}$; $\tau_1/m_1 = 3 \times 10^{14}$, $\tau_2/m_2 = 2 \times 10^{14}$. Antimony, $R\sigma = +5 \times 10^{-6}$ e.s.u., $B = 9 \times 10^{-11}$; $\tau_1/m_1 = 8 \times 10^{14}$, $\tau_2/m_2 = 4 \times 10^{14}$.

In the above discussion we have disregarded the fact that bismuth is a highly anisotropic metal, but, since the experimental data used refer to polycrystalline bismuth, this does not seriously affect the conclusions drawn. What we have discussed is a model isotropic metal which has the same properties as those of bismuth when the latter are averaged over all directions. Thus the values of τ/m calculated above should be the averages of the values of τ/m for the different directions calculated from a similar but anisotropic model. When we consider the values obtained for the various parameters, it is difficult to believe that they are in any way satisfactory, and we must conclude that the models used for multivalent metals have only a qualitative significance. The very small effective masses required lead to very large curvatures of the energy surfaces over a considerable energy range. There is some theoretical justification, based on a first order perturbation calculation, for believing that large curvatures can occur near an energy discontinuity, but, if we substitute the values found for n, m_1 and m_2 into the formula (1·13) for E_0, we find that the highest occupied energy level lies about 1 e.volt above the bottom of the upper band and about 0·4 e.volt below the top of the lower band; these energy ranges are too large for the levels to be considered as being in the neighbourhood of the energy discontinuity. Further, the more direct evidence of the conductivity of the bismuth alloys indicates that the number of conduction electrons is greater than that predicted by the indirect evidence, and on the whole the direct evidence is to be preferred.

6·4. The thermal conductivity.

It has been recognized for a very long time that the conduction of heat and of electricity in metals are closely related. As long ago as 1853 Wiedemann and Franz showed that κ/σ is very nearly the same for all metals at room temperatures (κ is the thermal conductivity), while in 1881 L. Lorenz showed that $\kappa/\sigma T$, the so-called Lorenz number, is very nearly constant over a large temperature range. It is therefore clear that most of the heat current is carried by electrons, although, as discussed in § 6·43, direct conduction by the lattice is also possible.

The calculation of the thermal conductivity is complicated by the fact that all thermal effects are second order effects which vanish for a completely degenerate electron gas, and thus the calculations have to be pushed much further than is the case, for example, with the electrical conductivity. For the thermal conductivity itself we can avoid the long calculations by using a result proved in books on kinetic theory with varying degrees of rigour (see, for example, Jeans, *Dynamical theory of gases*, chapter XII), which is valid for any single mechanism of conduction. It is

$$\kappa = \tfrac{1}{3} l \bar{v} C_v = \tfrac{1}{3} \tau \bar{v}^2 C_v, \tag{6·18}$$

where $\bar{v}$ is an average velocity and C_v is the specific heat (per unit volume) at constant volume. Detailed analysis is required in order to determine what particular average velocity is meant by $\bar{v}$. For metals, since only the fastest electrons take part in conduction, $\bar{v}$ is obviously the velocity of these electrons, and hence $\tfrac{1}{2} m \bar{v}^2 = E_0$, where E_0 is given by (1·13). Hence, by combining (6·18), (5·3) and (1·13), we obtain for the electronic thermal conductivity

$$\kappa = \frac{\pi^2}{3} \frac{n \tau k^2 T}{m}. \tag{6·19}$$

Further, from (6·2), we find for the Lorenz number L the expression

$$L = \frac{\kappa}{\sigma T} = \frac{\pi^2}{3} \left(\frac{k}{\epsilon}\right)^2. \tag{6·20}$$

6·41. The calculated value of L at 18° C. is $7·89 \times 10^{-11}$ e.s.u., which is in remarkable agreement with the observed value of $7·9 \times 10^{-11}$ e.s.u. obtained by averaging the results for twelve metals. At low temperatures, however, L differs considerably from the value L_0 given by (6·20), and thus both (6·19) and (6·20) are only valid at high temperatures. The reason is that in general it is impossible to define a time of relaxation uniquely and that the τ which occurs in (6·19) is not necessarily the same as the τ which occurs in (6·2). It is more or less a lucky accident that τ can be defined uniquely at high temperatures, and in all other cases we have to work with the general collision operator and not assume that it is of the simple form (6·1).

The exact dependence of κ on T has been found by Wilson (6); at high temperatures κ is constant, in accordance with (6·19), while at low temperatures it is proportional to T^{-2} for a perfectly pure metal. The behaviour of κ at intermediate temperatures is complicated and depends upon the number of electrons per atom. This is discussed in more detail in the next section.

6·42. The thermal resistance, like the electrical resistance, consists of a part due to impurities and an ideal part κ_i characteristic of the pure metal. There is no difficulty in defining a time of relaxation for collisions with foreign atoms and other fixed scattering centres, the free path being of the order of the distance between the centres. Thus for this part of the resistance the Wiedemann-Franz law holds, and, since the two scattering mechanisms are independent, the thermal resistances are additive and we have

$$\frac{1}{\kappa}=\frac{1}{L_0\sigma_0 T}+\frac{1}{\kappa_i}, \tag{6·21}$$

where $1/\sigma_0$ is the residual electrical resistance and $1/(L_0\sigma_0 T)$ is the residual thermal resistance.

In order to determine the ideal thermal resistance experimentally, the simplest procedure is to plot T/κ against T and thus to determine the limiting value of T/κ as T tends to zero. This gives us the residual thermal resistance and hence by subtraction the ideal resistance. A more indirect method was adopted by Grüneisen and Goens (7) based on "the law of isothermal lines". Grüneisen and Goens found that for different specimens of a metal at the same temperature the thermal resistance is a linear function of the electrical resistance, i.e.

$$\frac{1}{\kappa}=a+\frac{b}{\sigma}, \tag{6·22}$$

where a and b are functions of the temperature. By determining a and b experimentally, they found κ_i by substituting the known value of σ_i in (6·22). As we see by comparing (6·22) with (6·21) and (6·4), the functions a and b are given by

$$b=\frac{1}{L_0T}, \quad a+\frac{b}{\sigma_i}=\frac{1}{\kappa_i}. \tag{6·23}$$

The experimental results are too meagre and indefinite to make possible a rigorous test of the theoretical expression for κ_i. The results indicate that, for small T, $1/\kappa_i \propto T^3$ approximately. This is not in disagreement with the theory, because, although the theory predicts that $1/\kappa_i \propto T^2$ at very low temperatures, there is also a term in T^4 in the expression for $1/\kappa_i$, and thus if we put $1/\kappa_i \propto T^n$ the exponent n will lie between 2 and 4 for moderately low temperatures and it will only approach 2 for extremely low temperatures.

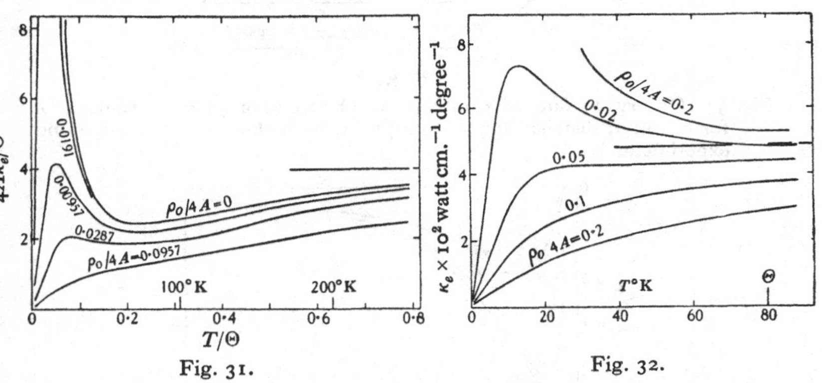

Fig. 31. Fig. 32.

Fig. 31. Theoretical electronic thermal conductivity for a monovalent metal, showing the effect of impurity. The temperatures marked correspond to copper (Θ=315° K.).

Fig. 32. κ_e for the bismuth model.

The theoretical electronic thermal conductivity κ_e has been computed by Makinson (8) taking into account the effect of the impurities. His results are shown in figs. 31 and 32; the parameter chosen to indicate the amount of impurity present is $\rho_0/4A$, where ρ_0 is the residual resistivity and A is calculated from the formula

$$\rho_i=AT/\Theta, \tag{6·24}$$

which is valid for high temperatures. For copper $A=1·82\times10^{-6}$ ohm, while for polycrystalline bismuth $A=3·3\times10^{-5}$ ohm.

For an ideally pure monovalent metal κ_e is constant at high temperatures, diminishes slightly at first as the temperature is lowered and then increases again as T^{-2} for very small T. When impurities are present, κ_e does not become infinite as T tends to zero, since, according to (6·21), $\kappa_e \propto T$ when the effect of the impurities is dominant. Hence, when the amount of impurity

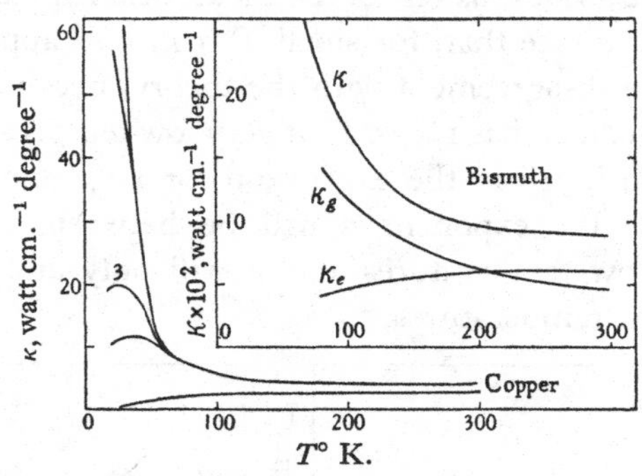

Fig. 33. Observed values of κ for copper. The trend of observed values of κ for bismuth, disregarding anisotropy. Curves show κ_e and κ_g as found experimentally.

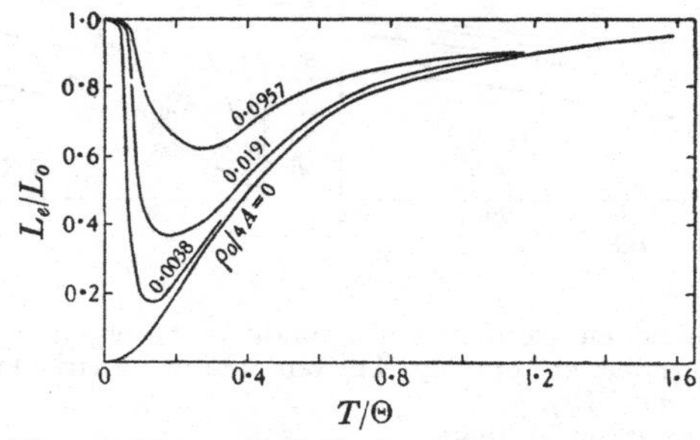

Fig. 34. The ratio L_e/L_0 for monovalent metals.

present is small, κ_e reaches a maximum at a very low temperature and then decreases to zero; the temperature at which κ_e is a maximum increases as the impurity content increases. When the amount of impurity is large, the maximum is flattened out and may disappear altogether. Some experimental results for copper are shown in fig. 33, the curve marked 3 being comparable with the curve $\rho_0/4A = 9{\cdot}57 \times 10^{-3}$ in fig. 31. The agreement is reasonable, and the only point in which there is any marked difference between theory and experiment is that the predicted flat minimum in κ_e is not observed.

The theoretical values of L_e/L_0 for a monovalent metal, where $L_e = \kappa_e/\sigma T$ and $L_0 = \frac{1}{3}(\pi k/\epsilon)^2$, are shown in fig. 34. For a perfectly pure metal L_e/L_0 decreases monotonically from 1 to 0 as the temperature decreases. For an impure metal, on the other hand, L_e/L_0 reaches a minimum and becomes 1 again when the effect of the impurities is dominant. These curves may be compared with the experimental curves shown in fig. 36.

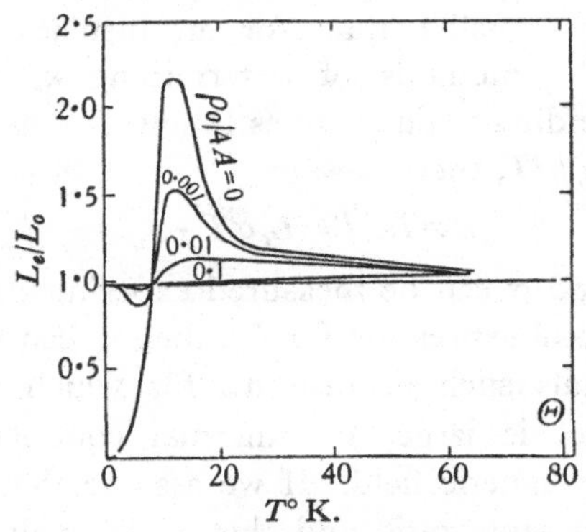

Fig. 35. The ratio L_e/L_0 for the bismuth model.

When the number of electrons is small, κ_e is of the form shown in fig. 32, which was constructed for a model isotropic metal which has the average properties of bismuth, it being assumed that the impurities are such that the number of electrons is the same for all specimens. The main difference between the results for this model and for a monovalent metal is that the flat minimum in κ_e no longer exists. The critical value of the number of conduction electrons per atom is 0·32 approximately. The ratio L_e/L_0 is shown in fig. 35.

6·43. The lattice conductivity.

Since insulators can conduct heat, it is clear that, though the electrons may be the most important carriers of heat in metals, yet the lattice vibrations must play some part in thermal conductivity. The electrons and the lattice vibrations provide two different mechanisms of heat transfer, and hence we can write

$$\kappa=\kappa_e+\kappa_g, \tag{6·25}$$

where κ_g is the lattice conductivity, the two heat currents being additive. This does not mean that the heat currents are independent of one another in the sense that we can add a lattice conductivity like that of an insulating crystal to the electronic conductivity to obtain the total conductivity. In a metal the free electrons provide a mechanism for scattering the lattice vibrations which is absent in an insulator, and hence κ_g for a metal must be smaller than for an insulator with similar properties. The methods of determining κ_e and κ_g must, therefore, be indirect. They are as follows:

(1) If $L_e=\kappa_e/\sigma T$, then

$$\kappa\equiv L\sigma T=L_e\sigma T+\kappa_g. \tag{6·26}$$

Now only κ and σ can be measured experimentally, but if we have a theoretical expression for L_e, then κ_g can be obtained.

(2) For metals such as bismuth for which the magneto-resistance effect is large, the thermal conductivity can be reduced by a magnetic field. If we assume that κ_e and σ are reduced in the same ratio and that κ_g is unaltered, then by measuring the change in κ and σ we can find κ_g.

In the experiments which have so far been carried out, it has been assumed that L_e has its high temperature value L_0, while, as figs. 34 and 35 show, this is by no means true at all temperatures. However, when $\rho_0/4A>0·01$, L_e/L_0 is nearly constant for bismuth, and this value of ρ_0 ($1·32\times10^{-6}$ ohm cm.) is exceeded in many specimens, so that serious error is not introduced by the assumption. Moreover, for the monovalent metals where L_e/L_0 differs widely from 1, the electronic conductivity is so large that the lattice conductivity is unimportant and is scarcely observable. However, if great accuracy is required, the only safe procedure is to calculate L_e using the appropriate values of $\rho_0/4A$ and of the number of electrons per atom. If the observed L is appreciably greater than L_e, then conduction by the lattice is of importance. Some experimental curves for L are shown in fig. 36.

The thermal conductivity of polycrystalline bismuth is shown in fig. 33 for a moderate temperature range; measurements at low temperatures by de Haas and Capel (9) indicate that κ reaches a maximum of the order of 1 watt cm.$^{-1}$ degree^{-1} near 19° K. Comparison of figs. 32 and 33 shows that the very high values of κ cannot be due to electronic conduction alone, the theoretical maximum value being much too low if we assume that there is a reasonable amount of impurity present. For bismuth, there-

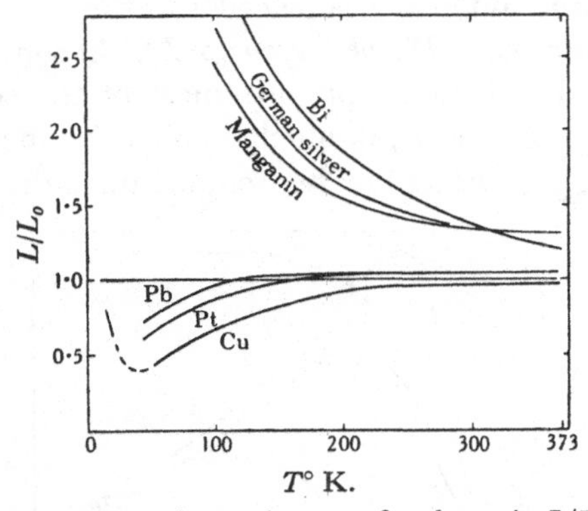

Fig. 36. Experimental curves for the ratio L/L_0.

fore, κ_g is considerable, and at low temperatures it is more important than κ_e. The values of κ_e and κ_g found experimentally are indicated in fig. 33. These were obtained by Grüneisen and Reddemann (10) by method (1); method (2) gave essentially the same results (11).

6·44. The thermal conductivity of the lattice is given by the same equation as is the conductivity of the electrons, namely

$$\kappa_g = \tfrac{1}{3} l u_0 C_v, \qquad (6{\cdot}27)$$

where l is the free path, u_0 is the velocity of sound and C_v is the specific heat of the lattice. In calculating the free path we have to take into account the following causes of scattering:

(1) interaction with the conduction electrons;

(2) irregularities of atomic dimensions in the lattice, i.e. irregularities small compared with the mean wave-length of the vibrations;

(3) lattice defects of large extent, such as grain boundaries;

(4) anharmonic coupling between the different lattice waves, which causes the waves to scatter one another.

With the exception of the first, all these causes of scattering are the same for metals and insulators.

The calculations have been carried out by Makinson (8); his conclusions are as follows. At very low temperatures the scattering is almost entirely due to grain boundaries. In this case l is constant and of the same order as the size of the crystal grains, and, since $C_v \propto T^3$, we have $\kappa_g \propto T^3$. At somewhat higher temperatures the electrons provide most of the scattering; in this region $\kappa_g \propto T^2$ and κ_g is much smaller for metals than for insulators (fig. 37). At still higher temperatures the mechanisms

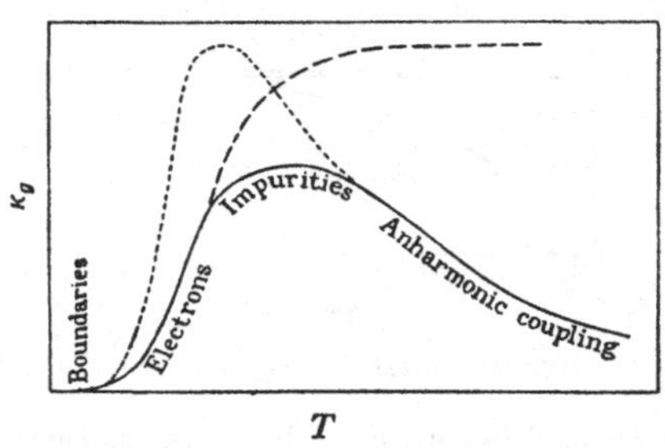

Fig. 37. The theoretical general form of κ_g. The dotted line shows the form for an insulator and the dashed line the form for a metal if only electrons scattered the lattice waves.

(2) and (4), which are mainly responsible for the thermal resistance in an insulator, are the most important; κ_g reaches a maximum and then decreases as T^{-1}.

The general form of the κ_g, T curve is shown in fig. 37, the main causes of scattering in each temperature range being marked. It is clear that κ_g can be important only for poor conductors like bismuth and then only for not too high temperatures. At high temperatures κ_e is constant while $\kappa_g \propto T^{-1}$, and so at sufficiently high temperatures $\kappa_e > \kappa_g$. For good conductors like copper not only is κ_e very large, except at the very lowest temperatures, but κ_g, on account of the large number of free electrons, is very much less than the conductivity of an insulator in the region where electronic scattering is important.

For bismuth, on the other hand, κ_e is small and κ_g is larger than in copper, so much so that κ_g is more important than κ_e at low temperatures.

The experimental data at present available are inadequate to test all the details of the theory, the experiments of de Haas and Capel (9) only just indicating that κ reaches a maximum. Recent experiments by de Haas and Biermasz (12) on quartz have, however, shown that boundary scattering occurs at very low temperatures. They found that instead of κ increasing as T^{-1}, as it should if the scattering is due to impurities distributed at random, κ reached a maximum at about 10° K., and that κ was nearly proportional to T^3 near 2° K. Since the maximum of κ for bismuth occurs at a much higher temperature, it seems fairly well established that the observed decrease in κ, and hence in κ_g, below 19° K. is due to the scattering of the lattice waves by the electrons.

6·5. The thermoelectric effects.

Since there is a lack of uniformity in the literature concerning the sign conventions for the thermoelectric effects, we begin by defining the quantities concerned. If wires of two dissimilar metals are joined at both ends and the two contacts are kept at

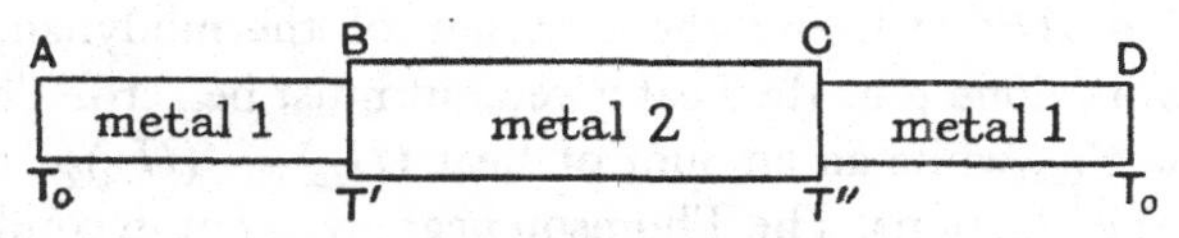

Fig. 38. The thermoelectric circuit.

different temperatures, an electromotive force is set up which manifests itself by producing a current. This electromotive force can be measured by opening the circuit at one point and measuring the potential difference. If the metals are arranged as in fig. 38, the thermoelectric force Θ_{12} is defined to be $V_A - V_D$. It is independent of T_0, and is a function of T' and T'' only. If T' is held fixed and $T'' = T$ is varied, then $d\Theta_{12}/dT$ is a function of T only. It is called the thermoelectric force per degree and is positive if $V_A - V_D$ is increased when T'' is increased. Alternatively, if $T' = T$ and $T'' = T + \Delta T$ ($\Delta T > 0$), then $d\Theta_{12}/dT$

is positive if the current flows from the hot to the cold junction in metal 1 when A and D are joined.

If an electric current passes from one metal to another, which is at the same temperature, heat is absorbed or emitted at the junction. The Peltier coefficient Π_{12} is defined as the heat given out per second when unit current passes from metal 1 to metal 2. Π_{12} is a function of the temperature of the junction.

When an electric current passes between two points of a homogeneous wire whose temperature difference is ΔT an amount of heat $\mu \Delta T$ per unit current is emitted or absorbed per second in addition to the ordinary Joule heat. The Thomson coefficient μ is taken as positive if heat is evolved when a positive current passes from the higher to the lower temperature. Both the Peltier and Thomson heats are proportional to the current strength and are thus reversible effects.

6·51. There exist well-known thermodynamic relations between the thermoelectric effects, which can be obtained as follows. Suppose that the points A and D in fig. 38 are joined by a wire of large resistance, so that any current passing round the closed circuit does so slowly enough for the process to be considered as reversible. Let unit charge pass round the circuit in the direction $ABCDA$. By the first law of thermodynamics the total work done plus the heat given out must be zero. The work done is Θ_{12}, while an amount of heat $(\Pi_{12})_{T'}-(\Pi_{12})_{T''}$ is given out at the junctions. The Thomson heat given out in conductor 2 is $-\int_{T'}^{T''} \mu_2 \, dT$, since the sign convention is such that this quantity must be positive if $T'>T''$; there is a similar expression for the Thomson heat in the conductor 1. Hence we have

$$\Theta_{12}+(\Pi_{12})_{T'}-(\Pi_{12})_{T''}+\int_{T'}^{T''}(\mu_1-\mu_2)\, dT=0. \quad (6·28)$$

Further, by the second law of thermodynamics, the total entropy change must be zero in this reversible process, which gives

$$\frac{(\Pi_{12})_{T'}}{T'}-\frac{(\Pi_{12})_{T''}}{T''}+\int_{T'}^{T''}\frac{\mu_1-\mu_2}{T}\, dT=0. \quad (6·29)$$

By differentiating (6·28) with respect to T'' and putting $T''=T$, we obtain

$$\frac{d\Theta_{12}}{dT}-\frac{d\Pi_{12}}{dT}+\mu_1-\mu_2=0, \tag{6·30}$$

and, by differentiating (6·29), we find

$$\frac{d}{dT}\left(\frac{\Pi_{12}}{T}\right)=\frac{\mu_1-\mu_2}{T}. \tag{6·31}$$

Eliminating $\mu_1-\mu_2$ from this and (6·30) we have

$$\Pi_{12}=T\frac{d\Theta_{12}}{dT}. \tag{6·32}$$

If we now substitute this expression for Π_{12} in (6·31) we obtain the remaining relation

$$T\frac{d^2\Theta_{12}}{dT^2}=\mu_1-\mu_2. \tag{6·33}$$

Of the three thermoelectric quantities only μ refers to a single metal. Borelius has, however, managed to obtain the absolute thermoelectric force of a single metal in the following way. We define the absolute thermoelectric force per degree $d\Theta/dT$ by

$$\frac{d\Theta}{dT}=\int_0^T\frac{\mu}{T}\,dT \tag{6·34}$$

in accordance with (6·33). In order to obtain $d\Theta/dT$ it is therefore necessary to measure μ for at least one metal down to the absolute zero. When this has been done, $d\Theta/dT$ can be found for other metals either by direct measurement of the relative thermoelectric power or by measuring the Thomson coefficients and integrating equation (6·33). The obvious plan of extrapolating the μ, T relation to the absolute zero is useless, since μ behaves in a very complicated way at low temperatures. Instead, use is made of the fact that there is no thermoelectric force between metals in the superconducting states, which is interpreted as meaning that $\mu=0$ for a superconductor. Hence by measuring the thermoelectric force of a metal against a superconductor we obtain the absolute thermoelectric force for temperatures below the transition temperature of the superconductor, and measurements of the Thomson coefficient enable us to obtain $d\Theta/dT$ at higher temperatures.

6·52. On account of the thermodynamic relations it is only necessary to consider one of the thermoelectric effects; the simplest to think of is the thermoelectric force. When one end of a homogeneous wire is heated, the electrons at that end have their energies increased and hence an electric current is set up. The heated end becomes positively charged, since electrons flow to the cold end until the electric field set up by the displacement of the charge is sufficient to counteract the effect of the temperature gradient. If the two ends are short-circuited by an ideal conductor whose properties are independent of the temperature, in fact by a superconductor, we see that there will be a current of electrons in the metal wire from the hot to the cold junction since the flow is now no longer hindered by the accumulation of charge. The absolute thermoelectric power is therefore negative. If, however, the current is due to holes in a nearly filled band, then the thermoelectric power is positive since the holes behave like positive electrons.

In order to obtain an idea as to the order of magnitude of the thermoelectric effects it is simplest to consider the Peltier effect. If we have an electric current, then at any point of a wire there is a flow of energy $n\bar{v}\bar{E}$, where n is the number of electrons per unit volume carrying the current, $\bar{v}$ is the average velocity and $\bar{E}$ is the average energy which each electron carries. Now if the current is a unit one we must have $n\epsilon\bar{v}=1$, and the energy flux is $\bar{E}/\epsilon$. We cannot find the average energy transported by the electrons without elaborate calculations, but we can find its order of magnitude quite easily. To do this, we must first be quite clear what the energy zero is. We are really interested only in the difference in the energy fluxes in two metals (the flux in a homogeneous metal is unobservable), and, since the Fermi energy ζ is the same at all points of a conductor in equilibrium, ζ is the obvious choice for the energy zero. $\bar{E}$ is then the mean thermal energy of an electron and it must be of the order of the specific heat per electron multiplied by the temperature. The fraction of the electrons which contribute effectively to the specific heat is of the order kT/E_0; these contribute $\frac{3}{2}k$, while the remainder contribute nothing. Therefore $\bar{E}$ is of the order k^2T^2/E_0, and the energy flux is of the order $k^2T^2/\epsilon E_0$.

Now consider two conductors in which a unit positive current flows from 1 to 2, the electronic current flowing from 2 to 1. At any point in 2 there is an energy flux of the order $k^2T^2/\epsilon(E_0)_2$ towards the junction, while in 1 there is an energy flux of the order $k^2T^2/\epsilon(E_0)_1$ away from the junction. The difference between these two energy fluxes is the Peltier heat given out at the junction. Hence Π_{12} is of the order

$$\frac{k^2T^2}{\epsilon}\left(\frac{1}{(E_0)_2}-\frac{1}{(E_0)_1}\right). \tag{6·35}$$

The numerical factor given by the calculations is π^2 for monovalent metals, and since $\Pi_{12}=T(d\Theta_1/dT-d\Theta_2/dT)$, we have (*T.M.* p. 177, equation (263) and p. 178, equation (265 (*a*)))

$$\frac{d\Theta}{dT}=-\frac{\pi^2k^2T}{\epsilon E_0} \tag{6·36}$$

and

$$\mu=-\frac{8\pi^2mk^2T}{\epsilon h^2}\left(\frac{\pi}{3n}\right)^{\frac{2}{3}}. \tag{6·37}$$

When the conductivity is due to holes, $-\epsilon$ must be replaced by ϵ.

In the preceding paragraphs we have been able to calculate the Peltier heat without in any way considering the mechanism by which it is produced. The reason for this is that owing to the definition of the Peltier coefficient we need consider only the current and not the electric field which produces it. Wherever there is a change of composition an electric field is set up of such a magnitude that the current is maintained constant throughout the conductor. It is this field of course which is responsible for the Peltier effect and the calculation of it lies outside the scope of the elementary theory. Since, however, we only require to know the amount of energy liberated, it is not necessary to calculate the field, but a knowledge of the field is required for the other thermoelectric effects. Here, however, the thermodynamic relations enable us to dispense with it once more.

6·53. The large thermoelectric effects in semi-conductors can readily be understood with the help of fig. 18, p. 59. We consider a couple formed by a metal and an excess semi-conductor, the positive direction of the current being from the metal. The electrons therefore are moving from the semi-conductor to the metal, and in the semi-conductor they have an energy $\frac{1}{2}\Delta E = kb$ relative to the Fermi energy ζ. Thus the energy flux per unit current is of the order kb/ϵ, and by the argument given in the preceding section, we have

$$\frac{d\Theta}{dT} \sim -\frac{kb}{\epsilon T}. \tag{6·38}$$

Similarly, if the semi-conductor is a defect conductor, the fastest electrons in the semi-conductor have an energy kb less than the Fermi energy ζ and so they absorb energy when passing into the metal. In this case the Peltier heat is negative, and the absolute thermoelectric force per degree is given by (6·38) but with $-\epsilon$ replaced by ϵ. The experimental results have been discussed in § 4·61.

6·54. We show in fig. 39 some typical results for the absolute thermoelectric force per degree at moderate temperatures. The behaviour at low temperatures is still more complicated, and

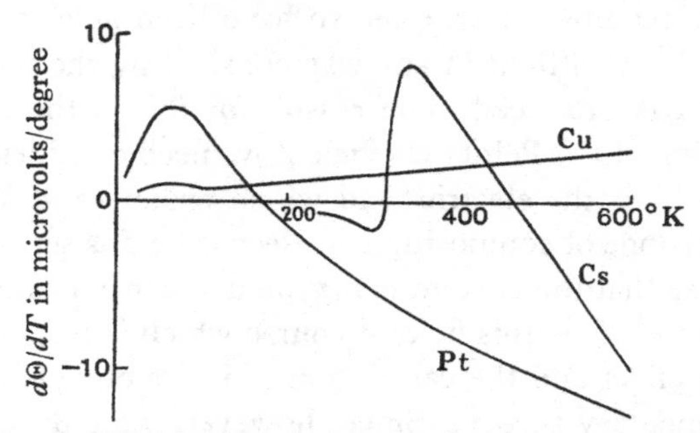

Fig. 39. The absolute thermoelectric force per degree as a function of the temperature.

cannot be explained at all by the theory in its present state. Now, since $d\Theta/dT$ is obtained by integrating μ/T (see (6·34)), any anomalies in μ at low temperatures persist in $d\Theta/dT$ at high temperatures, and hence it is a fairer test of the theory to compare the calculated and observed values of μ rather than of

$d\Theta/dT$. If then we find agreement for μ at high temperatures, any discrepancy in $d\Theta/dT$ is to be ascribed to the breakdown of the theory at low temperatures.

We should expect the theory to be reasonably correct for the monovalent metals, but this is not so. The Thomson coefficients of copper, silver, gold and lithium are all positive instead of negative at ordinary temperatures, while those of the other monovalent metals are indeed negative but have not the correct magnitude. This can only mean that the calculation of the thermoelectric effects requires an accurate knowledge of the energy levels and that the assumption that the electrons are free is not such a good one in this problem as it is, for example, in the calculation of the Hall effect. In some respects the theory is, however, fairly satisfactory. The proportionality of the Thomson coefficient to T and of the Peltier coefficient to T^2 is in good agreement with the experimental results for not too low temperatures. Further, if we assume that for Pd and Pt there is 0·5 electron per atom in the s-band and, on account of their low mobility, ignore the contribution of the holes in the d-band, the calculated values of μ/T are about -2×10^{-2} microvolt/degree for both metals. The observed values are $-3{\cdot}4\times10^{-2}$ for Pd and $-1{\cdot}8\times10^{-2}$ microvolt/degree for Pt. The smallness of the Hall coefficients (Table IV, p. 97), however, indicates that the holes in the d-band reduce the effect of the s-electrons, and so the agreement is not so good as it seems.

TABLE V. Thomson coefficients in microvolts/degree of the alkalis just above and just below the melting point

	Li	Na	K	Rb	Cs
μ/T:					
Solid	0·022	(0·005)	−0·036	−0·036	(0·008)
Liquid	0·030	−0·048	−0·043	−0·085	−0·076
Calculated	−0·016	−0·023	−0·036	−0·041	−0·048

One satisfactory piece of evidence is that the Thomson coefficients of the liquid alkali metals agree quite well with the values calculated on the assumption that the electrons are free. (See Table V. The experimental results, due to Bidwell(13), have been reduced to absolute values by Sommerfeld (14).) When a metal is in the liquid state the energy level system ought to be fairly simple, since the zone structure characteristic of the solid is lost, and this confirms us in ascribing the failure of the elementary theory, as applied to most metals, to our inadequate knowledge of the energy level system and not to the breakdown of the foundations of the theory.

REFERENCES

(1) J. Bardeen. *Phys. Rev.* **52** (1937), 488.
(2) N. F. Mott. *Proc. Roy. Soc.* A, **153** (1936), 699.
(3) N. Thompson. *Proc. Roy. Soc.* A, **155** (1936), 111.
(4) K. Ariyama. *Inst. Phys. Chem. Res.* (*Tokyo*), **34** (1938), 344.
(5) P. Kapitza. *Proc. Roy. Soc.* A, **123** (1929), 292.
(6) A. H. Wilson. *Proc. Cambridge Phil. Soc.* **33** (1937), 371.
(7) E. Grüneisen and E. Goens. *Zeit. Phys.* **44** (1927), 615.
(8) R. E. B. Makinson. *Proc. Cambridge Phil. Soc.* **34** (1938), 474.
(9) W. J. de Haas and W. H. Capel. *Physica* (2), **1** (1933), 929.
(10) E. Grüneisen and H. Reddemann. *Ann. Physik* (5), **20** (1934), 843.
(11) H. Reddemann. *Ann. Physik* (5), **20** (1934), 441.
(12) W. J. de Haas and T. Biermasz. *Physica* (2), **5** (1938), 320.
(13) E. C. Bidwell. *Phys. Rev.* **23** (1924), 357.
(14) A. Sommerfeld. *Naturwissenschaften*, **22** (1934), 49.

General reference

L. Campbell. *Galvanomagnetic and thermomagnetic effects.* (London, 1923.)

INDEX

INDEX